EXTENDING SCIENCE 5

METALS AND ALLOYS
Selected Topics

Extending Science Series

1	Air	E N Ramsden and R E Lee
2	Water	E N Ramsden and R E Lee
3	Diseases and Disorders	P T Bunyan
4	Sounds	J J Wellington
5	Metals and Alloys	E N Ramsden
6	Land and Soil	R E Lee
7	Energy	J J Wellington

Further titles are being planned, and the publishers would be grateful for suggestions from teachers.

EXTENDING SCIENCE

5

Metals and Alloys

Selected Topics

E N Ramsden BSc PhD DPhil

Stanley Thornes (Publishers) Ltd

First published 1985 by
Stanley Thornes (Publishers) Ltd
Old Station Drive
Leckhampton
CHELTENHAM   GL53 0DN

*An exception is made for the word puzzles on pp. 9, 44, 67, and 69. Teachers may photocopy a puzzle to save time for a pupil who would otherwise need to copy from his/her copy of the book. Teachers wishing to make multiple copies of a word puzzle for distribution to a class without individual copies of the book must apply to the publishers in the normal way.

British Library Cataloguing in Publication Data

Ramsden, E N
    Metals and alloys. — (Extending science; no. 5) .
    1. Metals
    I. Title   II. Series
    669   TN665

ISBN 0-85950-193-0

Typeset by Tech-Set, 15 Enterprise House, Team Valley, Tyne & Wear.
Printed and bound at Dotesios (Printers) Ltd, Bradford-on-Avon, Wiltshire.

CONTENTS

Chapter 1   What Use are Metals?

Spot the metal	1	How do we obtain metals?	3
What is a metal?	2	Questions on Chapter 1	8
What is an alloy?	3	Wordfinder on metals and alloys	9

Chapter 2   Copper, Bronze and Brass

The end of the Stone Age	10	Experiment 3 Weathering copper in a	
Bronze	12	hurry	16
The importance of copper today	13	Experiment 4 Etching copper	16
Experiment 1 The pretty green		Questions on Chapter 2	17
stones	15		
Experiment 2 An imitation of			
flotation	15		

Chapter 3   Gold

Gold in the Stone Age	18	Where to find gold	21
Gold in Ancient Egypt	18	Uses of gold	21
Gold in South America	19	Questions on Chapter 3	22
The gold rush	21		

Chapter 4   Silver

Native silver	23	Questions on Chapter 4	26
Uses of silver	23		

Chapter 5   Iron

The Iron Age	27	Corrosion	35
Iron swords and their teething troubles	28	Experiment 5 Investigating the	
Wrought iron and cast iron	30	strength of steel	38
Iron and the Industrial Revolution	31	Experiment 6 Zinc plating	38
The blast furnace	31	Questions on Chapter 5	39
Steel	32		

Chapter 6   Zinc

Uses of zinc	41	Crossword on iron and zinc	44
Questions on Chapter 6	43		

Chapter 7   **Lead**

Lead in ancient times   45
The extraction of lead   45
Uses of lead   45
Alloys of lead   48

Experiment 7 The lead–acid
    accumulator   48
Questions on Chapter 7   49

Chapter 8   **Tin**

The nature of tin   50
Canned food   50

Tin plague   51
Questions on Chapter 8   52

Chapter 9   **Mercury**

Mercury in ancient times   53
Mining mercury is no fun   54
Uses of mercury and its compounds   54

The nasty side of mercury   56
Questions on Chapter 9   58

Chapter 10   **Aluminium**

The newcomer   59
Uses of aluminium   61
Aluminium alloys   63
Aluminium and iron: partners in the
    car industry   65

Experiment 8 Anodising and dyeing
    aluminium   65
Questions on Chapter 10   66
Crossword on aluminium   67

Questions on metals and alloys   68

Metals and alloys wordsquare   69

PREFACE

The marvellous achievements of science and technology can make our lives safer and sweeter. The extraction and utilisation of metals is one of these achievements. Beautiful, strong and useful, metals find a host of diverse applications. I have tried to show how the uses to which each metal is put are determined by the properties of the metal.

The book is not intended to provide a complete coverage of metals. Many important subjects, such as the details of the methods used for the extraction of metals and the reactivity series of metals, have been omitted, on the assumption that pupils will be able to find this material in textbooks. Although I have provided a number of experiments, there are many others relevant to this topic. I know that pupils will be able to supplement the practical activities with the assistance of their teachers.

I hope that teachers will find the book useful and that their pupils will find it interesting. It may be used as extension material in a junior chemistry course or as part of a general science course. I hope that older pupils who find a traditional course too academic for their liking will find this material on the applications of science more to their taste.

E N Ramsden,
Wolfreton School, Hull

ACKNOWLEDGEMENTS

I am grateful to the colleagues and the consultant who have given me their time and their advice. They are Dr G H Davies, Mr J R Dennison, Mr R E Lee, Dr J H Lister, Mr B Rogers and Mr J Scott. I thank Miss Maxine Hamil for her contribution to the illustrations and Mr D F Manley for substantial improvements in the wordfinder and crosswords. A considerable amount of design goes into the production of a topic book, and I thank the publishers for the care which they have taken over this volume. My family have helped me as usual, with their interest and encouragement.

The author and publishers are grateful to the following who provided photographs and gave permission for reproduction.

Barnaby's Picture Library (p. 2);
Paul Brierley (p. 25 and main cover photograph);
British Aerospace (p. 64 upper);
British Museum (p. 13);
British Steel Corporation (pp. 32, 33, 34 upper and cover photograph);
Capper Pass (p. 34 lower);
Griffith Institute, Ashmolean Museum, Oxford (p. 18);
Guy's Hospital Dental School (p. 56);
Popperfoto (p. 62);
Royal Institute of British Architects (p. 46);
RTZ Photographic Library (pp. 4, 5, 37 and 41).

WHAT USE ARE METALS?

SPOT THE METAL

Many of the objects shown below are made from metals and alloys. Can you say which metal is used for each object? If not, try again on p. 68.

WHAT IS A METAL?

There are 92 elements that occur on the Earth's surface. (An element is a substance that cannot be broken down into simpler substances.) Of these elements, 70 are metals. We find uses for about half of these. The rest of the elements are non-metallic elements, such as sulphur and oxygen. They have important uses.

How can we tell a metallic element from a non-metallic element? What are metals like? A scientific way of asking this question is to say, 'What are the *properties* of metals?'

Some properties of metals

1) Metals are solids. The exception is mercury, which is a liquid at room temperature.

2) Metals are shiny. Some metals become dull or tarnished when exposed to the air. A freshly cut surface of any metal is shiny. Gold is a metal that never loses its shine. Sodium is a metal that becomes tarnished in minutes.

3) Metals are strong. They can be bent without breaking. They are described as *malleable,* which means that they can be hammered into different shapes. They are *ductile,* which means that they can be pulled out to form wire.

Metals are malleable

4) Metals are good conductors of heat. This is why saucepans and frying pans are made of copper or steel or aluminium.

5) Metals are good conductors of electricity. This property is the best way of telling whether a substance is a metal. One non-metallic element, however, is also a good conductor of electricity. This is a form of the element carbon called *graphite*.

What use do we make of metals?

Different metals have different properties. Copper is a very good *electrical conductor,* and is used for electrical wiring. Aluminium has a low density. Boats made from aluminium are lighter than steel boats. Silver and gold are so beautiful that they are used in jewellery.

WHAT IS AN ALLOY?

Sometimes, there is no metal ideally suited for a particular purpose. In this case, chemists combine two or more metals to form an alloy. An alloy is a metallic substance. It has different properties from the metals of which it is composed. Chemists have to match up the properties of an alloy with the job that it has to do. The chemists who specialise in preparing metals and alloys are called *metallurgists.*

Solder is an alloy of lead and tin. It melts at a lower temperature than either tin or lead. Steel is an alloy of iron with other metals and the non-metallic element, carbon. When a soldier fires a machine gun repeatedly, the barrel becomes very hot. If it becomes so hot that the metal softens, the gun will not fire straight. The alloy used for making gun barrels must have a high melting temperature. Metallurgists added the metal molybdenum to steel to make an alloy suitable for this purpose.

HOW DO WE OBTAIN METALS?

We obtain metals from *minerals.* Any naturally occurring substance which is not of plant or animal origin is called a mineral. Sand and rock are minerals. Some minerals contain enough metal to make it worth while extracting. Such a mineral is called an *ore.* Obtaining metals from their ores is a long and costly business. There are two main ways of obtaining metals from their ores:

Prospecting

This is the search for deposits of metal ores. Prospectors must have a good knowledge of geology and be able to recognise minerals. A prospector will often crush a sample of ore and put it in a miner's pan. Then he swishes it round with water. While soil and debris wash over the rim of the pan, the denser minerals sink to the bottom. Panning can give a rough idea of the value of an ore. If the ore looks promising, it is sent to a laboratory. There a chemist analyses it.

Panning for gold

Modern prospectors have scientific techniques to help them. They can look at photographs taken by satellites as they circle the Earth. To a trained eye, these show where the mineral deposits may be found.

Mining. (a) Drilling to extend the Wheal Jar tin mine in Cornwall

(b) Tipping ore into the underground crusher

Mining

Some different types of mine are shown below.

(a) A strip mine or open-cast mine. The ore lies on the surface or under a thin layer of soil. The surface is removed in strips. Aluminium ore is mined this way

(b) A shaft mine. The ore is buried below a layer of rock. A vertical shaft is sunk into the ground. Tunnels lead from the shaft to the deposits

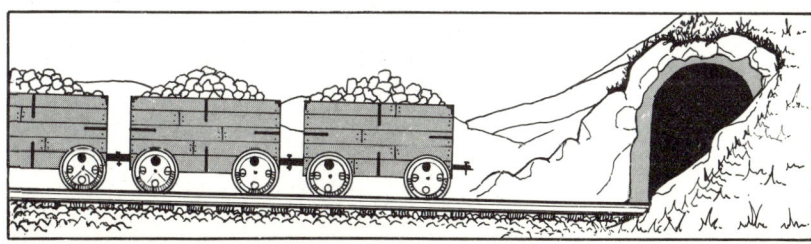

(c) A portal mine. A sloping tunnel is driven into the side of the mountain

Treatment of ores

The diagram below shows how the ore is treated after it has been mined.

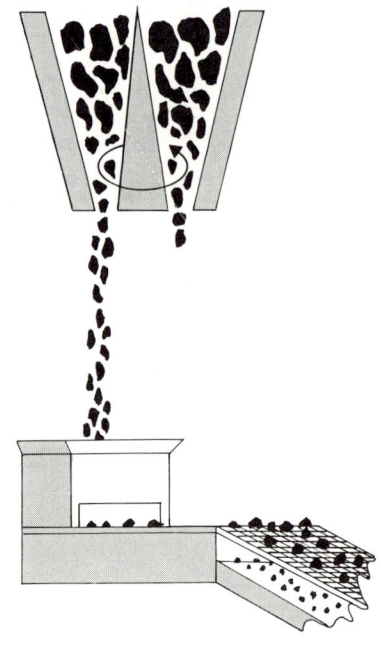

1
Crushing A cone crusher is shown here. The smaller cone spins round and round. Rocks are broken up between the two cones. Small pieces can pass through the opening at the bottom

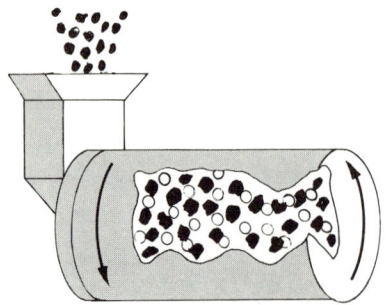

2
Screening (sieving)
The screens (sieves) are set in motion. Small lumps fall through

3
Grinding One method of grinding is to use a ball mill. Steel balls tumble around inside a rotating cylinder. The lumps of ore are crushed to powder between the moving balls

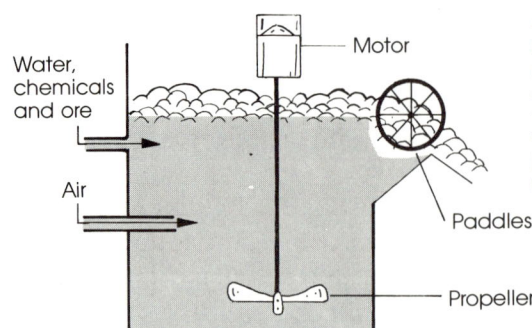

Water, chemicals and ore

Motor

Air

Paddles

Propeller

4
Concentrating the ore The propeller stirs the mixture. An oily foam is formed. Mineral particles stick to it to form a slime. Paddles scrape off the mineral-carrying slime. It is dried and sent to the smelter. Rubbish (called **gangue**) is left behind in the tank. This method is called **froth flotation**

Smelting of ores

Smelting is a common method of obtaining metals from their ores. Before being smelted, ores are roasted. Many of the ores are sulphides.(A sulphide is a compound of the metal with sulphur.) When sulphides are roasted, sulphur is removed as the gas sulphur dioxide. The metal is converted into its oxide (a compound of the metal and oxygen). For example:

Zinc sulphide + Oxygen → Zinc oxide + Sulphur dioxide

Next, the oxide is smelted. Chemically speaking, this means *reducing* the oxide (taking away oxygen) to give the metal. It is done by heating the oxide with coke, an impure form of carbon, in a furnace. The gases carbon monoxide and carbon dioxide are formed. The metal is left behind. For example:

Zinc oxide + Carbon → Zinc + Oxides of carbon

Sometimes, a substance must be added to combine with the impurities. The aim is to turn them into a molten *slag*. Molten slag floats on top of molten metal at the bottom of the furnace. Both are run off from time to time through taps.

Break-even point

All these operations cost money. For the combined processes, the value of the metal obtained must be greater than the expense of obtaining it. There is a possibility that, as it goes through all these stages, the mineral will reach a *break-even point*. This means that the cost of extracting the metal is equal to the price that can be obtained for it. No one can make a profit this way. The ore is abandoned. The prospector has to look for another ore.

Why are different methods used for the extraction of different metals?

Some metals are described as *reactive;* others as *unreactive*. What we mean by a reactive metal is one that is ready to take part in chemical reactions. It is ready to combine with other elements to form compounds. Aluminium is a very reactive metal. As soon as it comes into contact with air, it reacts. It combines with oxygen in the air. The surface of the metal becomes covered with a film of aluminium oxide. Gold is a very unreactive metal. A gold ring can be worn for years, exposed to air and water and chemicals. It never becomes tarnished.

We say that an unreactive metal like gold 'occurs *native*' (which means that it occurs in nature as the free element). It has been in the Earth for millions of years without combining with any other element. Reactive elements, such as aluminium and iron, are not found free. Both aluminium and iron react with air and water. They could not exist on the Earth's surface without reacting. They are found as compounds. The more reactive an element is, the more difficult it is to extract it from its compounds. Aluminium is more reactive than iron. It is much more difficult to obtain aluminium from aluminium oxide than to obtain iron from iron oxide. Lead and copper are rather unreactive elements. It is fairly easy to obtain these metals from their compounds. The method of extraction which is used for any metal depends on how difficult it is to persuade that metal to let go of the elements with which it is combined. That is to say, it depends on how reactive the metal is.

Scrap metal for recycling

QUESTIONS ON CHAPTER 1

1 Explain the meanings of these terms:
(a) element, (b) metallic element, (c) property, (d) alloy, (e) tarnish, (f) corrosion, (g) density, (h) reactive.

2 Explain why aeroplanes and guns and wedding rings are not all made of the same metal.

3 What is the difference between a metal and an alloy? What advantages do alloys have over metals?

4 Metal oxides can be split up into the metal and oxygen. Mercury oxide splits up at 200°C. Iron oxides split up when heated to 1000°C with coke. Aluminium oxide can only be split up by electricity. What does this information tell you about the reactivity of these three metals? Which is the most reactive? Which is the least reactive of the metals?

WORDFINDER ON METALS AND ALLOYS

First, trace the wordfinder on to a piece of paper (or photocopy this page — teacher, please see the note at the front of the book). Then solve the following clues and put a ring around each answer. Answers go in any direction: across, back, up, down and diagonally. When you have ringed all the answers, you will have been a successful miner and the unringed letters taken in order (left to right, row by row) will spell out a message. The answer to Question 1 is ringed already to give you a start.

P	W	F	S	O	L	D	E	R	G	E	L
A	R	L	O	P	E	N	C	A	S	T	L
N	C	O	R	N	W	A	L	L	O	I	M
G	D	T	S	O	I	S	N	N	D	N	U
A	E	A	Y	P	R	O	E	U	I	D	N
N	H	T	A	V	E	V	N	E	U	N	E
G	R	I	N	D	E	C	A	C	M	O	D
U	D	O	I	K	A	S	T	C	O	B	B
E	V	N	A	E	L	I	I	O	R	R	Y
E	L	E	A	D	L	D	V	T	R	A	L
H	R	E	G	E	O	O	E	L	D	C	O
B	I	R	O	N	Y	R	U	C	R	E	M

1 A valuable mineral may sink to the bottom of it (3)
2 He looks for valuable minerals (10)
3 If a metal is ____ (7), it can be drawn out . . .
4 . . . into a ____ (4)
5 This metal is a liquid at room temperature (7)
6 This metal helps make gun barrels strong (10)
7 This metal becomes tarnished very quickly (6)
8 The surface of an ____-____ mine is removed in strips (4-4)
9 At the ____-____ point the cost of extracting a metal is equal to the price you will get for it (5-4)
10 Froth ____ is used in the treatment of ores (9)
11 A ball mill is used to ____ an ore (5)
12 An unreactive metal occurs '____' (6)
13 Molten ____ must be separated from molten metal (4)
14 The rubbish left behind when an ore is concentrated (6)
15 A combination of two or more metals (5)
16 An example of 15 (6)
17 16 contains this metal (4) . . .
18 . . . and this one (3)
19 Steel contains this metal (4) . . .
20 . . . and this non-metallic element (6)
21 The Wheal Jane tin mine is in this county (8)

2

COPPER, BRONZE AND BRASS

THE END OF THE STONE AGE

This chapter begins in 5000 BC, during the period called the Stone Age. The only tools and weapons which our ancestors had at this time were made from stones. Among the stones, people began to find native metals. You remember from Chapter 1 that these are metals occurring 'free' in nature and not combined with other elements. They include copper, silver and gold. When a rock containing a native metal was put into a very hot camp-fire, the metal melted. In the ashes of the fire, a nugget of the metal would be found. Our ancestors used the shiny metals to make bracelets, pendants and other ornaments.

They found that copper was useful as well as ornamental. It is harder than silver and gold, but it is soft enough to be hammered into shape with a stone hammer. During the hammering process, copper becomes harder. The Stone Age hunter was used to chipping away slowly and painstakingly at stones to make arrowheads, spearheads and knives. Can you imagine his frustration if the stone cracked as he was nearing the end? To hammer away at a piece of bright, shiny copper, knowing that he could not smash it, was a great improvement.

(a)   (b)

(a) Chipping a stone

(b) Hammering copper

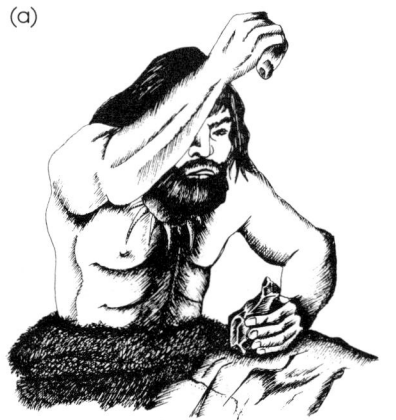
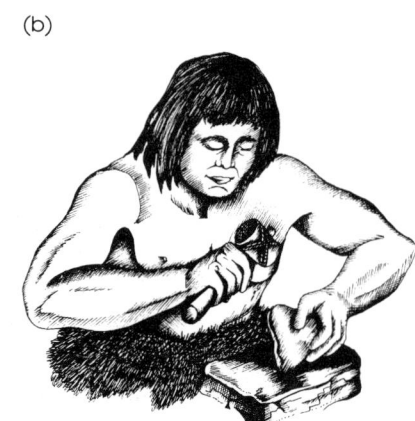

As well as occurring native, copper is found in copper compounds. In time, people discovered how to obtain copper from ores containing copper compounds. Imagine this scene in about 3500 BC. A Stone Age woman is building a hearth to hold the family's fire. She chooses some pretty green stones. When the children return home with armfuls of firewood, she lights the fire. After a while, the wood burns to form charcoal. By the time the hunter of the family brings home his kill, there is a glowing red charcoal barbecue waiting for him. Next day among the ashes, the woman finds a lump of shiny, reddish gold copper. It looks to her as though the heat of the fire has melted the green stones, and molten copper has run out to solidify on cooling. In reality, what happened was not melting; it was *smelting*. Smelting is the extraction of a metal from its ore.

A Stone Age fireplace

The domestic fireplace was not ideal for smelting copper ores. People learned by experiment to construct a furnace over a hole in the ground so that molten metal would collect in the hole. A good breeze blowing through the furnace helped to raise the temperature. At some time, bellows, made out of animal hides, were invented, as shown in the illustration on p. 12. Next, people learned to line the hole in the ground with clay. This absorbed some of the soil and ashes. Then a removable lining was invented. If the lining could be lifted out of the furnace, the metal in it could be poured into moulds.

A primitive furnace

Among the most useful articles which could be moulded were copper cooking pans. Before this time, bowls and dishes were made of pottery. Copper pans were so much stronger and could be heated so much more quickly that our ancestors must have been delighted with them.

BRONZE

The batches of impure copper produced by the early smelters were not always the same. Metal made from some rocks was harder and therefore made better tools than copper from other sources. The reason was that many copper ores also contain tin compounds. Smelting had produced an alloy of copper and tin called bronze. Bronze is harder than copper and can be ground to a sharper edge. Both these characteristics made bronze tools better than copper tools.

Bronze weapons led to a revolution in hunting. Bronze tools speeded up every aspect of farming. They made it easier to clear the land. They made it easier to plough and to sow and to harvest crops. The Stone Age gave way to the Bronze Age. For the first time, hunting and farming no longer occupied the time of every member of a community. People had time to develop other skills. Different people in the community did different jobs. There were farmers, potters, painters, soldiers and so on. By 3500 BC, several parts of the world were 'civilised'.

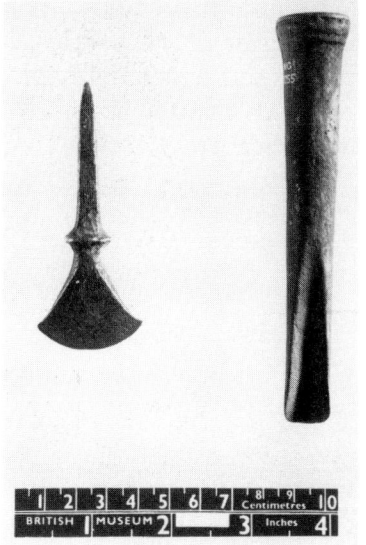

Bronze Age tools

THE IMPORTANCE OF COPPER TODAY

The world production of copper is 8 million tonnes a year.
Copper is mined as copper sulphide, which miners call *copper
pyrites*. The treatment of the ore is described in Chapter 1.

Uses for copper and its alloys

Copper is not attacked by water. It can therefore be used
for roofing important buildings. It reacts slowly with air to
form a surface layer of green copper compounds. These
compounds are basic copper carbonate and basic copper
sulphate. You will have noticed how some buildings have
attractive green roofs. This is because they have been
roofed with copper and the copper has 'weathered'.

A copper dome and
a bronze statue

Copper is a good conductor of heat. It has been used for centuries for cooking pans and kettles. Nowadays, aluminium pans and stainless steel pans are more popular. They are cheaper than copper pans, and they do not become tarnished. (See pp. 35 and 61–62.)

A copper kettle

Copper is a good electrical conductor. It is also soft enough to be drawn out into wire. Copper wire is used in electrical circuits. Very pure copper is needed for this purpose. A single building may contain several miles of copper wire. Electrical industries take more than half of the world's production of copper. Wires, cables, overhead power lines, switches and windings in electric motors are made of copper. The transatlantic telegraph system began in 1866 when 3000 miles of copper wire were laid under the Atlantic.

Copper has been used for a long time for the manufacture of water pipes and hot water tanks. Being an unreactive metal, it is not attacked by water. Also, it is easy to weld.

Small quantities of other metals are often added to make alloys which are harder than copper. Bronze (copper and tin) is made into statues, medals, gun metal, springs and coins. Bronze makes a pleasing sound when it is struck. It is *sonorous*. For this reason, it is used for making church bells. It is also used for organ pipes.

14

Brass is an alloy of copper with zinc. Since it can be polished to a brilliant yellow colour, it is used for making ornaments. It is easy to work, easy to cast, and easy to machine. Musical instruments such as trumpets are made of brass. It is used for making engineering fittings, such as taps and screws, and for making ships' propellers.

Other copper alloys are used for the manufacture of coins. Coinage metals must be hard and able to stand up to hard wear. Coins must carry a pattern. The more complicated the pattern, the more difficult it is for a counterfeiter to copy the coin. During manufacture, presses stamp down on to the coins. They leave a pattern imprinted on the coins. Coinage metals must be soft enough to take on the pattern of the presses.

Here are some suggestions for experiments on copper. You will need some help from the teacher with Experiment 1, and Experiment 3 is a teacher demonstration.

EXPERIMENT 1

The pretty green stones

Ask your teacher for some of the green copper compound which the cave woman admired. Ask your teacher to suggest some tests which you can do on the compound. Try to find out what the substance is.

EXPERIMENT 2

An imitation of flotation

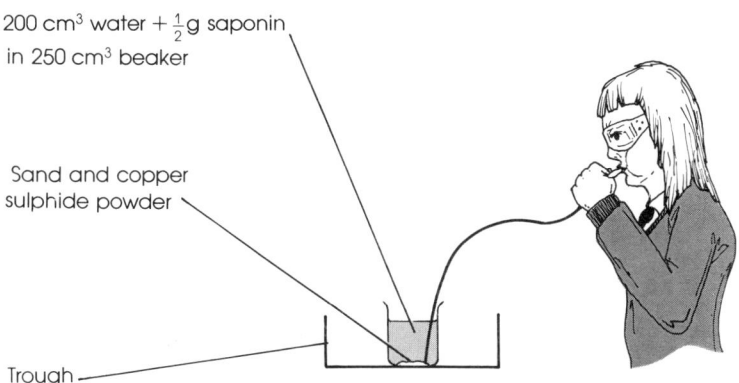

200 cm³ water + ½ g saponin in 250 cm³ beaker

Sand and copper sulphide powder

Trough

1) Blow air through a piece of rubber tubing for 5 minutes.
2) Direct the air on to the bottom of the beaker so that it will stir the mixture. The froth formed will overflow into the trough.
3) Next, filter the contents of the trough through filter paper. Examine the residue in the filter paper. Compare your results with p. 70.

Weathering copper in a hurry: TEACHER DEMONSTRATION

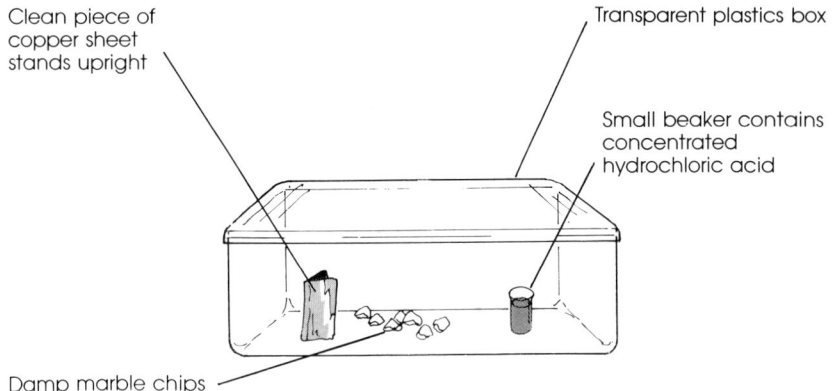

Clean piece of copper sheet stands upright

Transparent plastics box

Small beaker contains concentrated hydrochloric acid

Damp marble chips

Set up the apparatus shown in the diagram above. Watch carefully to see what happens to the copper sheet.

Etching copper

The technique of *etching* copper is important. Etching means removing material by some chemical means. The reagent iron(III) chloride reacts with copper. It is used to etch decorative patterns on copper bracelets and brooches. More importantly, it is used to make printed electrical circuits. Circuit board is a plastics material covered with a thin layer of copper. Iron(III) chloride will etch right through the layer of copper on a circuit board.

For this experiment, you need a copper sheet or circuit board, a candle and iron(III) chloride solution.

1) Take a piece of copper sheet or circuit board 5 cm square.

2) Light a candle, and cover the sheet with wax.

3) Cut a pattern in the wax. You must expose the copper below.

4) Put the sheet into a solution of iron(III) chloride. Leave it for 20 minutes.

5) Remove the sheet. Rinse it with water. Chip away the wax.

6) If you are working with circuit board, you can try covering patches of circuit board with Sellotape. Then spray the board with varnish. Remove the Sellotape. Dip the board into iron(III) chloride solution. The area which is not protected by varnish will be etched away. The varnish has to be removed with a solvent.

When people use circuit board to print electrical circuits, they use a method similar to this. They use a special pencil instead of Sellotape. The pencil deposits a layer of wax on the board. The protected area eventually becomes the electrical circuit.

QUESTIONS ON CHAPTER 2

1 What is meant by the statement 'copper occurs native'? Imagine that one Bronze Age man is telling another how to make a copper arrowhead, starting from an ore containing native copper. What instructions might he give?

2 Bronze gave its name to a new age, the Bronze Age. This shows the importance which historians attach to the discovery of bronze.

   (a) What is the difference between copper and bronze?

   (b) Why were bronze tools better than copper tools?

   (c) Before the Bronze Age, people had only stone and pottery from which to make their possessions. What differences did bronze make to their lives?

3 Name another alloy of copper, and list its uses.

4 The uses to which copper is put depend on its physical and chemical properties. List four uses of copper. Explain what chemical or physical property of copper makes it suitable for each use you have listed.

5 Supply words or phrases to fill the blanks in this passage. Do not write on this page.

Copper sulphide ore is crushed. This is done by ____. Then the small lumps of ore are separated from larger lumps. This is done by ____. The small lumps are ground in a machine which ____. The powdered ore is concentrated. Particles of ore are separated from rubbish, called ____, by the method of ____. The way this is done is ____. Next, the ore is roasted because roasting ____. After roasting, copper ____ is obtained. From this, copper is obtained by ____.

GOLD

GOLD IN THE STONE AGE

Gold was one of the first elements known to humans. It occurs native. Stone Age people loved the gleam of gold-bearing rocks. When they heated such rocks strongly, gold trickled out. They caught the trickle of metal in a mould. Being softer than copper, gold was not used for tool making. It was prized for its beautiful colour and shine. From the earliest times, people have loved to wear gold ornaments. There is no sign of a change in this habit. Gold never becomes tarnished.

GOLD IN ANCIENT EGYPT

The Egyptians were already skilful goldsmiths in 3500 BC. They extracted gold from rocks by breaking up the rocks with stone hammers and then grinding them. A stream of water washed away rock particles and left the denser gold particles behind. These particles were gathered up and melted and poured into moulds. A boy called Tutankhamun was King of Egypt in 1350 BC. His tomb was discovered in 1922. The mummy of the king was enclosed in three coffins.

Tutankhamun's death mask

The two outer coffins were overlaid with gold. The innermost coffin was pure gold. Goldsmiths had worked elaborate patterns in it. The king's death mask was of gold.

GOLD IN SOUTH AMERICA

South America is rich in gold. The Incas of Peru dominated an empire which reached the length of the Andes mountains. The empire was at its height at the time when Columbus discovered the Americas.

The Inca Empire in relation to present-day countries

The Andes reach heights of three thousand metres. At such heights, the nights are freezingly cold. The morning Sun bringing warmth and light each day is a welcome sight. The Incas worshipped the Sun. They built a marvellous Temple of the Sun God in their capital city, Cuzco. They decorated it in the colour of the Sun, in gold. The walls of the temple were covered from top to bottom with gold. At the eastern end, a solid gold Sun with solid gold rays was framed with gems. One of the temple courts contained a golden garden. Golden figures of animals stood between gold and silver bushes and trees.

19

The Incas obtained most of their gold from river beds. They collected particles of gold by panning. They melted the gold and poured it into moulds to make ingots. Goldsmiths fashioned articles from ingots. They warmed the metal and hammered it into shape. The smiths liked to work with alloys. They found that alloys are easier to melt than pure gold. By alloying copper with gold, they obtained a reddish-gold coloured metal. By alloying with silver, they obtained a pale yellow gold or a white gold alloy.

Inca goldwork

The Spaniards invaded Peru in 1532. They are called the Spanish *conquistadores* (conquerors) because they captured so much of South America. The Inca emperor, Atahuallpa, offered them hospitality. The Spaniards valued both his gold and his empire. They took him captive and killed six thousand of his troops. Although the Spanish party was made up of only 60 cavalry and 150 infantry, it had advantages over the Incas. The Spaniards had horses and they had guns. The Incas had bronze axes and spears. It was a contest between the Bronze Age and the Iron Age, and the Iron Age won. The emperor, Atahuallpa, was imprisoned. The Spaniards offered him for ransom. His subjects collected the largest ransom the world has ever seen. The treasure weighed 24 tonnes and was worth £15 million (at today's prices). After collecting the largest ransom in history, the conquistadores executed their prisoner.

After the largest ransom in history

THE GOLD RUSH

There was a gold rush in Australia in 1851. Gold was discovered in Victoria and in Melbourne. Thousands of prospectors flocked there. By 1853, there were sixty thousand
gold miners. The gold was not difficult to find. After
digging down 2 or 3 metres, the miners reached 'wash dirt'.
Sometimes, they were able to pick lumps of gold out of this
with a knife. The rest they 'panned' in a stream. Soil washed
out of the pan, and the particles of gold stayed in the pan.
A man could often mine 2 or 3 kilograms (4–6 lb) of gold in
a morning. One famous single nugget was worth £10 000. It
was named the 'Welcome Stranger' nugget.

WHERE TO FIND GOLD

Three-quarters of the world's gold is mined in South Africa.
It is found as small flecks of gold, scattered through the
rocks.

There is plenty of gold in the sea. Seawater contains many
dissolved salts. If we could extract all the gold from the gold
salts in seawater, we would have ten thousand million
tonnes of gold. This would be enough to fill a million
swimming pools. People have tried methods of obtaining
gold from seawater. So far, the cost of the process has been
greater than the market price of the gold extracted. No
such process has reached a *break-even point* (see Chapter 1).

USES OF GOLD

Gold is an international currency. Gold coins have been
used for centuries. Now, countries keep their gold in the
bank vaults, and print paper money instead. Most of the
gold produced in the world goes into the gold reserves
which every country keeps.

A good deal of gold is used in making jewellery. Besides
being a beautiful metal, it is easily worked. It can be
hammered into thin sheets of gold leaf. These can be so
fine as to be almost transparent. From a cupful of gold can
be obtained enough gold leaf to cover four football pitches.
Gold leaf is used for gold lettering and for gilding book
edges and picture frames.

Alloys of gold are used. The other metal is often copper.
The gold content is measured in *carats*. Pure gold is called
24 carat gold. Eighteen carat gold is 18/24 pure gold, that is
75% gold.

Other metals are often plated with gold. This can be done by an electrical method. Gold-plated contacts are used in electrical circuits for special purposes.

Rolled gold is made by welding a plate of gold to another metal and rolling it to the required thickness. The film of gold may be up to a quarter of a millimetre thick and does not wear off easily. Rolled gold is used for watch cases, cuff links, the frames of glasses and some jewellery.

Gold has medical and scientific uses. Dentists use it to fill teeth and to cap teeth. It is so unreactive that there is no danger of gold reacting with any of the things we eat and drink. Doctors sometimes treat cancer patients with radioactive elements. The radiation which these elements give off destroys cancers. There is a radioactive type of gold. Tiny grains of radioactive gold can be fired into the patient's body. This puts them in the right place to do their job. At close quarters, the radioactivity destroys the cancer. Healthy tissues are not damaged. Being perfectly harmless, the gold can be left inside the body.

QUESTIONS ON CHAPTER 3

1 Explain why gold was one of the first metals to be used by the human race. Why was gold not used for making weapons?

2 Describe how the Egyptians obtained their gold in ancient times. Why do we know so much about the work of the Egyptian goldsmiths?

3 Describe how the Incas obtained their gold. What pleasure did it give them? Would you say that gold was a blessing or a curse to the Incas?

4 List four uses for gold. Explain why gold is the metal chosen for each of these uses.

5 Supply words or phrases to fill the blanks in this passage. Do not write on this page.

Gold is used for making jewellery because it is ____ and ____ and ____. Alloys of gold are also used. An alloy is ____. The gold content of an alloy is measured in ____. What this means is ____. There are two ways of plating another metal with gold. One is ____. The other is ____.

SILVER

NATIVE SILVER

Silver is found native in Mexico and Argentina. In fact, the name Argentina means 'land of silver'. In Australia, South America, the USA and Canada, it is found as silver sulphide.

USES OF SILVER

Silver is used for making jewellery. It is a soft metal which is easily worked. Silversmiths can fashion the metal into intricate designs.

A drawback of silver is that it tarnishes readily. City air contains hydrogen sulphide. This gas is formed when sulphur-containing fuels burn. Silver reacts with it slowly to form silver sulphide, which is black.

In the state of New Mexico, USA, the American Indians prize their silver jewellery. The Navaho, the Hopi and the Zuni tribes are skilful silversmiths. Each tribe has traditional designs, which are passed from one generation to the next. Turquoise and ivory and coral are worked into the designs. Indian families often do not bank their money. They prefer to buy silver and turquoise. The family's 'bank balance' may be worn around the mother's neck in the form of a turquoise and silver necklace.

Navaho turquoise and silverwork

Silver and silver-plated metals have always been popular for tableware. People like silver teapots, sugar bowls and cream jugs. Silver is used for trophies, cups and shields. Silver-plating is done by electroplating.

Silver articles carry a *hallmark,* as you can see in the illustration below.

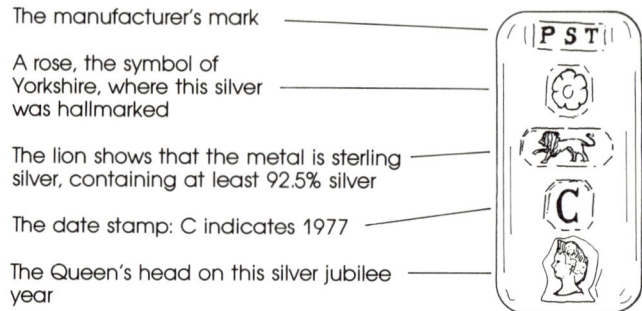

The manufacturer's mark

A rose, the symbol of Yorkshire, where this silver was hallmarked

The lion shows that the metal is sterling silver, containing at least 92.5% silver

The date stamp: C indicates 1977

The Queen's head on this silver jubilee year

Many nations have used silver coins. The UK has not used any silver in coins since 1946. Alloys are used instead. When the USA used silver dollars instead of paper money, prospecting for silver was a popular occupation. In Nevada there is a 'ghost town' called Calico. It sprang up practically overnight in 1859. News of a local find brought prospectors flocking. The silver streak soon ran out, and the miners moved on, leaving Calico as you see it.

Calico today

Nothing else reflects light as well and as evenly as silver. Reflected light gives silver its pale white sheen. Even the thinnest sheet of silver will reflect 95% of the light falling on it. Silver is used to coat mirrors. Recently a new use has been found for silver mirrors. They are being used to harness energy from the Sun. Banks of huge, curved silver mirrors focus the Sun's rays on to *photocells.* These convert light energy into electricity.

Solar collectors

Silver is a good conductor of electricity. Small discs of silver make contact in electrical switches in telephones, computers, dishwashers and other electrical appliances. Silver contacts open and close with little friction, and therefore little heat is generated. Silver is used as a lubricant. It is plated on the bearings of jet engines and diesel engines. Silver seals are used in the engines of the space shuttle to reduce friction, because in this engine any friction that might give rise to sparks must be avoided. Sparks could set off an explosion of the shuttle's fuel.

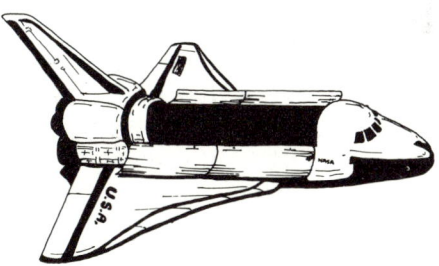

Space shuttle

Silver is used to purify drinking water. In the presence of silver, oxygen will kill bacteria. Silver is not used up in the process, but it has to be present for the process to occur. It is what chemists call a *catalyst*. This method of purifying water is not widespread, but it is used by airlines. It avoids the taste of chlorinated water. Swimming pools can also be disinfected in this way. The water is pumped through charcoal filters containing silver. There is no need to add chlorine, with its irritating effect on swimmers' eyes. The number of people who can afford to give their swimming pools the silver treatment is very limited!

QUESTIONS ON CHAPTER 4

1 There are two reasons why silver is used for making jewellery. What are they? What disadvantage is there in silver jewellery?

2 List articles in your home and school which are made of silver or plated with silver. What is sterling silver? How can you tell the difference between sterling silver and silver plate?

3 Some years ago, silver was needed in large quantities for a certain purpose. Silver is no longer used for this purpose. What is it? New uses for silver have been found. Mention one new and important use.

4 Explain the importance of silver in (a) electrical circuits, (b) the space shuttle, (c) swimming pools.

CHAPTER 5

IRON

THE IRON AGE

The Bronze Age was followed by the Iron Age. The skills of Bronze Age people in working copper ores were adapted to iron. A higher furnace temperature was needed for smelting iron ores. Iron has a melting temperature of 1535 °C, and would not melt in a primitive furnace. The iron obtained would look more like a porous stone than a metal. While it was still hot, a lump of iron could be beaten into shape with a stone. During the shaping, impurities were beaten out of the iron. In this way, early people made iron tools. By 1200 BC, iron had become an important metal.

A primitive furnace for smelting iron ores

Exhaust gases

Iron ore and charcoal

Iron forms

Charcoal

Clay lining

Air from bellows

The iron produced in these primitive furnaces was of the type later called *wrought iron*. At first, it was no improvement on bronze. It was soft and would not take a sharp edge and make a sword. Sometimes, the iron was left in the furnace longer than usual or the furnace was unusually hot. Then some carbon might combine with the iron. If this happened, a harder kind of iron would be formed. The harder metal was more useful for making tools and weapons. Sometimes, the product could be very brittle. This is the kind of iron we now call *cast iron*.

You can imagine that a primitive smelter might drop a piece of hot iron into a pool of water to cool. When he picked it up, he would find that the soft metal had become hard and brittle. No longer could it be hammered with a stone without breaking, but it could now be ground to make a sword or knife which was much stronger than any he had used before. The smelter had discovered the effect of rapid cooling which we call *quenching*. If he heated the metal and then allowed it to cool slowly, a less brittle metal resulted. It could be bent and worked without breaking. This is the treatment we call *annealing*. The best iron was made by *tempering*. In this process, quenching (rapid cooling) was followed by heating and slow cooling. Not only was the tempered iron hard, it could also be worked without breaking.

IRON SWORDS AND THEIR TEETHING TROUBLES

The quality of iron produced by the early smelters was very variable. A strong batch of iron was superior to bronze. It took a sharper cutting edge and made better tools and weapons. Iron swords could be unreliable, and the early smelters never knew what had gone wrong if a batch of iron was not up to standard. The Gauls were defeated by the Romans at a battle near Milan in 223 BC. The long iron swords of the Gauls were easily bent. After one mighty blow, the edges turned and the blade bent. A warrior had to straighten the sword with his foot, against the ground, before he could deliver a second blow. He was likely to get a Roman legionnaire's sword through his ribs while he was doing this. A thousand years later, the Vikings were still having the same kind of trouble.

(a)

(b)

(a) A Bronze Age warrior

(b) A Viking swordsman

The early metallurgists could not explain why some batches of iron turned out stronger than others. When by chance a good piece of metal *was* obtained, and turned into a mighty sword, it was believed to have a magical origin. King Arthur led his Knights of the Round Table into battle with his magical sword Excalibur. A legend describes how King Arthur came to own his sword. Excalibur was given to him by the Lady of the Lake. She had spent nine years making it, toiling in her cavern beneath the waters of the lake.

Was Excalibur really magical? A sword must be hard, and it must be flexible. These are two properties which are very difficult to achieve at the same time. It can be done. When iron is heated and then cooled quickly, layers of flexible pure iron and hard iron–carbon alloy are built up. Modern metallurgists have electron microscopes to help them in working out the methods of doing this. The primitive metallurgists used the method of trial and error. Sometimes, they succeeded in combining hardness and flexibility. Then, a sword that they made would take a sharp cutting edge, and would not break or bend when it struck armour plate. When they succeeded, it must have seemed like magic.

By the fourteenth century, the Japanese had become very good at making swords. They were able to achieve the combination of hardness and flexibility. The smiths hammered out the metal into a thin layer, folded it and hammered it out again. They repeated the process until the metal was ten or twenty layers thick. Then they heated the metal and *quenched* it (cooled it rapidly). The result was a metal that contained thousands of layers of flexible pure iron and thousands of layers of hard iron–carbon alloy. The Samurai were the military class, who were entitled to wear arms.

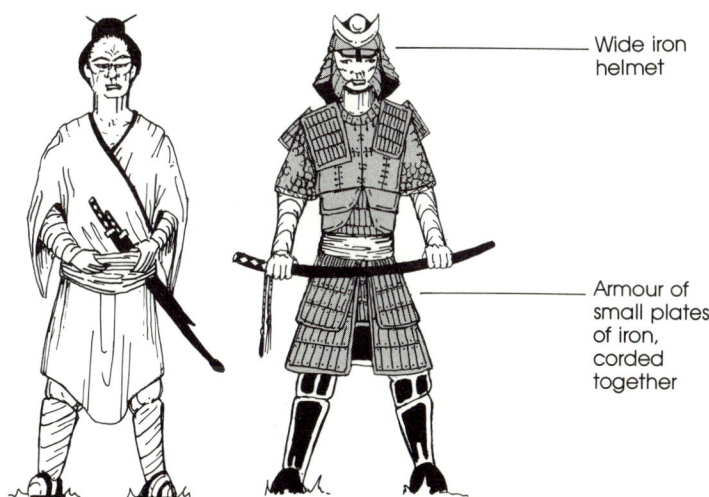

Samurai swordsmen

Wide iron helmet

Armour of small plates of iron, corded together

They paid several hundred pounds for a sword from a famous sword maker. They liked to take the precaution of testing their swords before staking their lives on the metallurgist's skill. A Samurai warrior might bribe the public executioner to use the warrior's sword to execute a prisoner. Once the swordsman had seen his sword cut off a head, he felt much happier about trusting his own life to it.

WROUGHT IRON AND CAST IRON

The earliest furnaces produced wrought iron. Wrought iron is iron containing less than 0.25% carbon. While it is hot, it can be hammered and rolled into shape without breaking.

Blacksmiths traditionally used wrought iron for making horseshoes. Ornamental ironwork was made from wrought iron. When iron has to be bent and worked into a complicated design, wrought iron is needed.

Wrought iron objects

As larger and larger furnaces were built, higher temperatures were reached. At high temperatures, carbon combines with iron, and cast iron is formed. Cast iron contains between 3 and 4% carbon. It is brittle and cannot be worked like wrought iron. The carbon content gives it a lower melting temperature than pure iron. Cast iron can be melted and allowed to set in moulds (that is *cast*) more easily than wrought iron. Cast iron expands slightly on cooling. This makes it flow into all the corners of a mould. It reproduces the shape of a mould exactly. By casting, objects with complicated shapes can be made. These would be difficult to make in any other way.

Cast iron objects

IRON AND THE INDUSTRIAL REVOLUTION

The Industrial Revolution took place in the UK between 1780 and 1860. Iron was an essential part of the revolution. It was needed for the construction of machinery. These machines produced in quantity articles that had previously been made by hand. The nineteenth century saw revolutionary improvements in transport. The railways were built. Iron was needed for the manufacture of the trains and the railway lines. The production of cast iron in the UK rose from 30 000 tonnes in 1760 to 90 000 tonnes in 1800. Today, it is 1000 million tonnes a year.

Iron is the basis of our technology in the twentieth century. Our cars and trucks, trains and ships are made of steel. Our manufacturing industry needs machines made of steel. Our buildings are built around a framework of steel girders.

THE BLAST FURNACE

A modern blast furnace for making iron costs about £1 million, and stands about 50 metres high. It is a tower made of steel plates and lined with heat-resisting bricks. A blast of hot air is sent in near the bottom. Iron ore and coke and limestone are fed in at the top. There is plenty of limestone in the UK and most other countries. Coke is obtained by heating coal in the absence of air. Many countries have plentiful supplies of coal. Iron ores are widespread over the surface of the Earth.

A number of chemical reactions take place in the blast furnace. As a result, molten iron trickles to the bottom of the furnace. Also formed is a molten mixture of compounds called a *slag*. The process runs continuously. A blast furnace runs for years without shutting down.

A blast furnace

STEEL

The iron that comes out of the blast furnace is cast iron. The high carbon content makes it too brittle for many purposes. Fortunately, carbon can be burnt off as its oxides. These are carbon monoxide and carbon dioxide, which are both gases. Calcium oxide (quicklime) is added to remove impurities. It combines with them to form a molten slag. The iron which is left behind contains about 1% carbon. Iron which contains this small percentage of carbon is called *steel*.

In 1850, Henry Bessemer invented a daring new 'converter' for turning iron into steel.

The difficulties were enormous. If the air pressure was too great, molten metal was blown out of the container and sprayed around. If the air pressure was too low, molten iron would run back into the air holes and block them. It was no wonder that there were 'teething troubles'. But Henry Bessemer was not the sort of person to say, 'It can't be done'. He overcame the difficulties, and in 1856 the process began to run successfully. Suddenly the price of steel dropped to one-fifth of its former value.

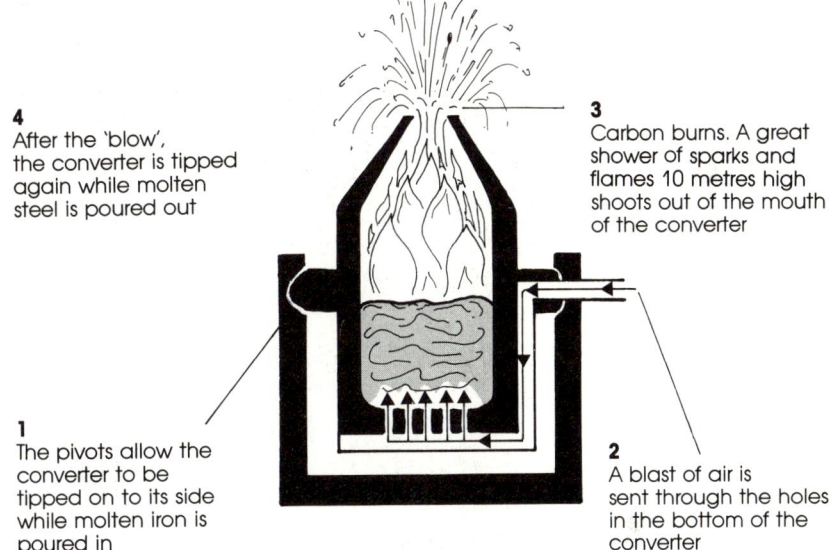

The Bessemer converter

4 After the 'blow', the converter is tipped again while molten steel is poured out

3 Carbon burns. A great shower of sparks and flames 10 metres high shoots out of the mouth of the converter

1 The pivots allow the converter to be tipped on to its side while molten iron is poured in

2 A blast of air is sent through the holes in the bottom of the converter

Improvements have been made to Bessemer's process. The steel industry now uses oxygen instead of air for oxidising the carbon present in cast iron. The steel made by the oxygen process is better than Bessemer steel. It contains no nitrogen. This is an advantage because nitrogen makes steel brittle.

Modern steelmaking — filling a converter with molten iron

Taking a sample of
steel from a converter

The control room of a
modern steel plant

Types of steel

Alloy steels contain metals other than iron. Different metals give the steel different properties. Some of the metals which are alloyed with iron are listed below. All steels contain carbon.

Different steels

Element	Type of steel	Uses
Carbon $\frac{1}{4}$%	Mild steel	Food cans, buckets, chains
Carbon $\frac{1}{2}$%	Medium-carbon steel, tough and springy	Car springs and axles
Carbon 1–4%	High-carbon steel, hard and brittle	Files, axes, saws, razor blades
Nickel	Increases strength and resists attack by acids	Tools, machinery, acid-resisting steels to be used in, e.g., road tankers for carrying sulphuric acid
Chromium and nickel	Stainless steel	Cutlery, car accessories, especially parts of cars where rusting is likely, e.g. exhaust pipes
Titanium	Strong even at high temperatures Low density	A titanium steel was designed especially for Concorde. Concorde flies at 2400 kilometres an hour. The air friction raises the temperature of the fuselage to 150 °C. The metal used for construction must not soften at this temperature. It must also be low in density
Tungsten	Hard and tough, even at high temperatures	High speed cutting tools. It is important that such tools keep a sharp cutting edge, even at high temperature. Propeller shafts

CORROSION

The corrosion of metals is a nuisance. The cost to the UK of replacing corroded metals is £500 million a year. Iron and steel rust when exposed to damp air. A trace of acid in the air speeds up the process. In industrial areas, factories burn coal and oil. These fuels contain sulphur. Sulphur burns to form the acid gas sulphur dioxide (see *Extending Science 1: Air*). When sulphur dioxide is present in the air, rusting takes place much faster.

For many families, the cost of corrosion is the cost of replacing the family car. The bodywork rusts before the engine wears out. Salt is spread on the roads in the winter to melt the ice. When salt reaches the bodywork of cars it speeds up the rusting process. To reduce rusting as much as possible, car manufacturers paint the body during manufacture. If the paintwork is scratched, the body rusts underneath.

Rustproofing a car body

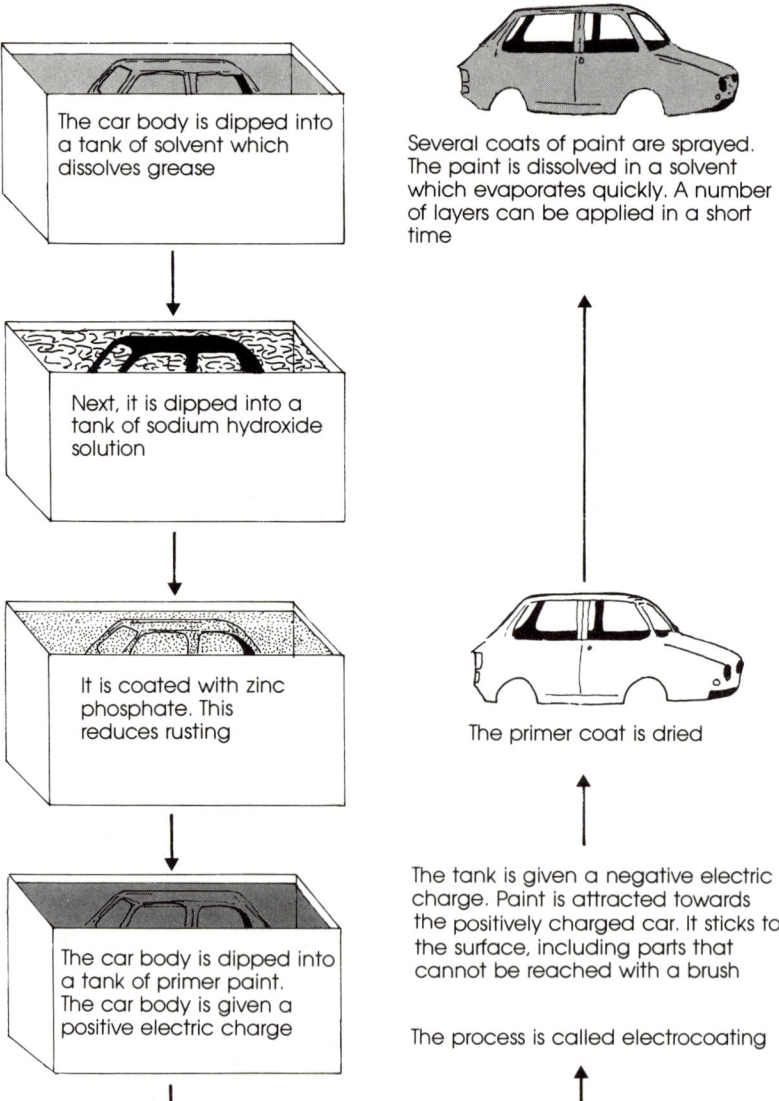

The car body is dipped into a tank of solvent which dissolves grease

Next, it is dipped into a tank of sodium hydroxide solution

It is coated with zinc phosphate. This reduces rusting

The car body is dipped into a tank of primer paint. The car body is given a positive electric charge

Several coats of paint are sprayed. The paint is dissolved in a solvent which evaporates quickly. A number of layers can be applied in a short time

The primer coat is dried

The tank is given a negative electric charge. Paint is attracted towards the positively charged car. It sticks to the surface, including parts that cannot be reached with a brush

The process is called electrocoating

Some parts of the car, such as the bumpers and the radiator, are more likely to be damaged than the rest. They are not painted: they are given a coating of chromium. This is applied by electroplating. A layer of nickel is electroplated first. This is very resistant to corrosion. Then a layer of chromium is added. It is harder than nickel, and, with its shiny appearance, more attractive.

Chromium plating

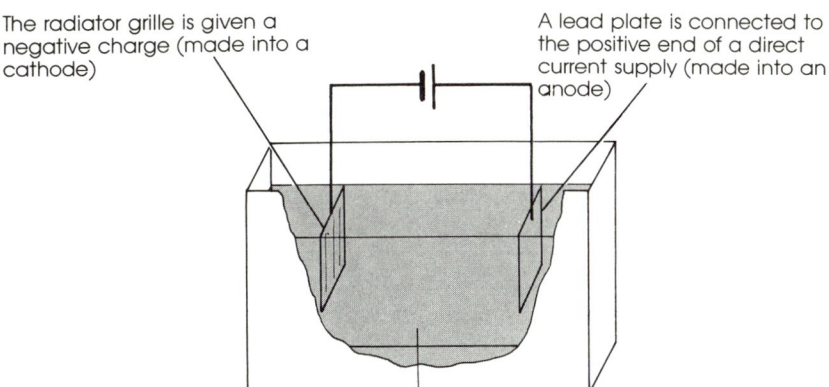

The radiator grille is given a negative charge (made into a cathode)

A lead plate is connected to the positive end of a direct current supply (made into an anode)

The bath contains a solution of chromium sulphate. Chromium sulphate contains chromium ions. These are positively charged particles of chromium. They travel to the negatively charged grille. There they lose their charge to become atoms of chromium. A layer of chromium forms on the grille

Road bridges and railway bridges can become corroded quickly. Exhaust gases from vehicles and trains contain sulphur dioxide. In damp air, this leads to the formation of sulphuric acid, which eats into metals. Galvanised steel is often used for bridges. Galvanising means coating with a layer of zinc. You can read about zinc in Chapter 6.

Part of a car-ferry terminal at Holyhead, with steel girders protected by galvanising

Investigating the strength of steel

1) Hold a steel hairpin with tongs. Heat it to a bright red heat.

2) Immediately, dip the hairpin into cold water. This is called *quenching*.

3) You will see that the tough, springy nature of steel is replaced by brittle hardness. The steel is easily snapped.

4) Take a second hairpin. Keep it red hot for $\frac{1}{2}$ minute. Let it cool slowly. It will bend easily. This is the process of *annealing*.

5) The softened steel obtained in (4) can be *tempered*. First, treat it as in (1) and (2). Then heat it until it becomes a blue colour. When it cools, it becomes tough and springy. Most steels are worked in this way. They lose their brittleness and keep their toughness.

Zinc plating

1) Take a heavy iron nail, and clean it as follows. (a) Sandpaper it. (b) Let it stand in sodium hydroxide solution for 2 minutes. (c) Drop it into hydrochloric acid. (d) Wash it in water.

2) Set up the apparatus shown below.

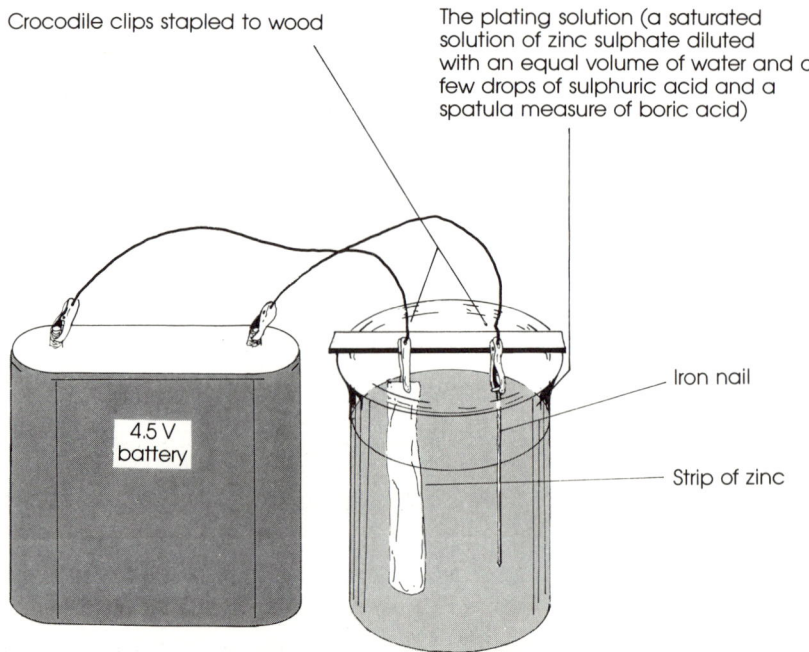

Crocodile clips stapled to wood

The plating solution (a saturated solution of zinc sulphate diluted with an equal volume of water and a few drops of sulphuric acid and a spatula measure of boric acid)

4.5 V battery

Iron nail

Strip of zinc

3) Allow a current to pass for 5 minutes.

4) Take out the galvanised (zinc-plated) nail. Drop it into a test tube full of sodium chloride solution.

5) Put an untreated nail into another test tube of sodium chloride solution. Label both test tubes.

6) Compare the two nails after two days and then after a week. (Compare your results with p. 70.)

QUESTIONS ON CHAPTER 5

1 (a) Why did iron smelters need a different kind of furnace from that which they used for copper ores?

(b) How did the ironsmith make iron into arrow-heads etc?

(c) What is meant by *quenching?* What does it do to iron?

(d) Why did the quality of iron obtained by primitive smelters vary from one batch to another?

(e) Why did the type of iron which the smelters obtained change gradually over the years from wrought iron to cast iron?

2 Why did the need for iron suddenly increase in the nineteenth century? Give examples of the purposes for which iron was used.

3 What raw materials are used in a blast furnace? How are these materials obtained? Why do you think that blast furnaces were built in South Wales and not in Cornwall; in West Yorkshire and not in Cheshire? A look at an atlas will help you to answer.

4 What is a *converter?* How does it work? Why do modern converters use oxygen instead of air?

5 Describe how the manufacturers try to prevent cars from rusting. What treatment is given to the body-work? How are the bumpers treated? Why do the bumpers need extra protection?

6 Imagine that you have to get through the day without using any articles made of iron or steel. List the things you normally use which you would have to do without.

7 Why do you think that we use more steel than any other metal? What useful properties does it have? Why are cars made of steel, rather than copper or lead or tin or aluminium? What unfortunate property does steel have?

8 The parts of a car are made from different metals. Explain why (a) body panels are made of steel, not cast iron, (b) the engine is made of cast iron, (c) medium-carbon steel is used for axles and springs, (d) chromium steel is used for ball bearings, (e) chromium-plated steel is used for door handles.

CHAPTER 6 ZINC

USES OF ZINC

Zinc is mined as zinc sulphide. (See Chapter 1, p. 7.) The world consumption of zinc is $4\frac{1}{2}$ million tonnes a year. Brass-making uses 20% of the zinc produced. (For brass, see Chapter 2.) A large fraction (40%) of the zinc produced is used to form protective coatings for iron and steel. (See Chapter 5.) This is very important because protection from rusting can save so much money every year.

Using zinc to prevent iron from rusting

The reason why a coat of zinc protects iron is that zinc is the more reactive of the two metals. Even if the coating of zinc is scratched, the iron underneath does not rust. There are five methods of applying a zinc coating.

Hot dip galvanising is one method of applying a zinc coat. The object is dipped into a bath of molten zinc. As the object is lifted out, a layer of zinc sticks to the surface. Zinc is chemically bonded to the iron beneath.

Hot dip galvanising

For large objects, *spraying* is more convenient. A nozzle directs a spray of molten zinc on to the surface. It will not reach inside tubes and other awkward places.

Electroplating is used for small objects. It gives a finer finish than galvanising. This method is used for instruments where rough and uneven finishes cannot be tolerated. Nuts and bolts and screws will keep a thread if they are *electro-galvanised*.

Small articles are often coated by *sherardising*, a process illustrated below.

Sherardising

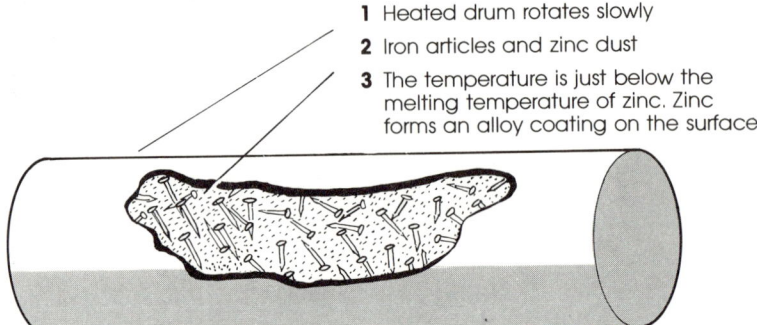

1 Heated drum rotates slowly

2 Iron articles and zinc dust

3 The temperature is just below the melting temperature of zinc. Zinc forms an alloy coating on the surface

Zinc dust paints consist of very fine zinc dust suspended in a liquid. The paint dries to give a film containing 90% zinc. Such paints are often called zinc-rich paints. The illustration below shows how different parts of a Land Rover are protected by zinc.

Zinc and the Land Rover

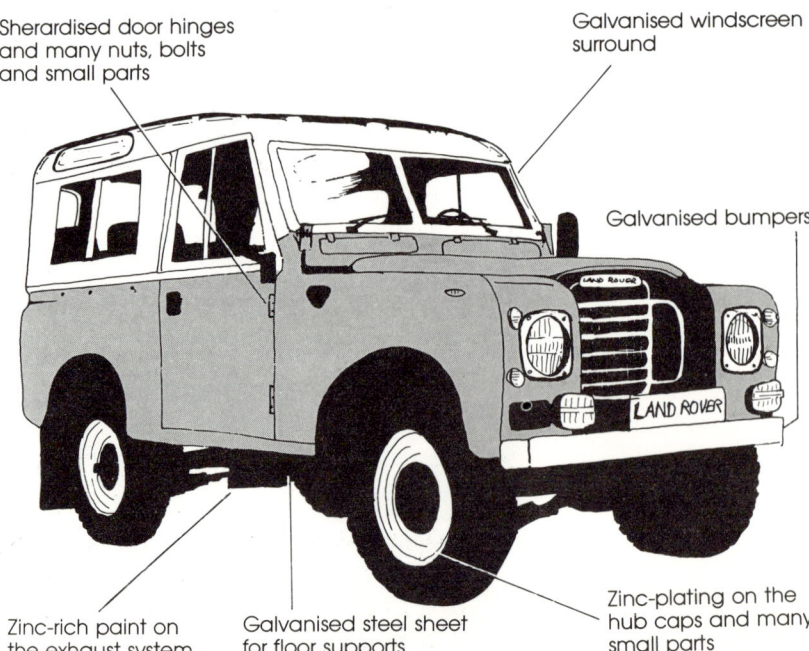

Sherardised door hinges and many nuts, bolts and small parts

Galvanised windscreen surround

Galvanised bumpers

Zinc-plating on the hub caps and many small parts

Galvanised steel sheet for floor supports

Zinc-rich paint on the exhaust system

Zinc is used to protect ships from rusting. If a lump of zinc is attached to the hull of a ship, the zinc will become corroded. This means that the steel hull will not rust as long as there is any zinc left. A time will come when the zinc needs to be renewed. This method of protecting steel is called *sacrificial protection.* The lumps of zinc are sacrificed to protect the steel. If the skipper does not remember to replace the zinc from time to time, he is in for a nasty surprise!

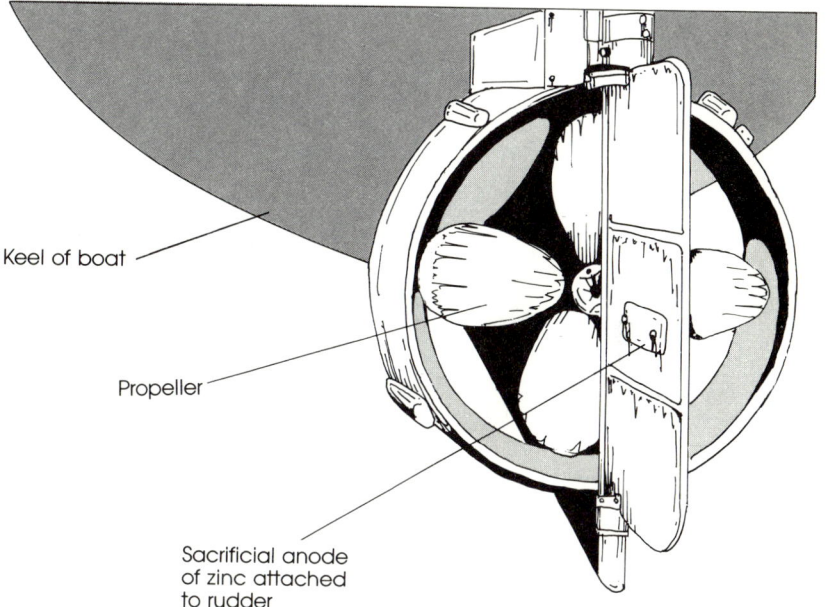

Sacrificial protection

Keel of boat

Propeller

Sacrificial anode
of zinc attached
to rudder

QUESTIONS ON CHAPTER 6

1 Which of the methods, dipping, spraying, sherardising or electroplating, would be most suitable for (a) a dustbin, (b) a knife, (c) a car bumper, (d) wire mesh fencing, (e) nails, and (f) steel tubing? Explain your choices.

2 Before being sherardised, iron must be dry. If water is present, it will react with zinc. What will be formed? Why will this be dangerous? Why is it especially dangerous in the sherardising method? (This is a question for those who know their chemistry. See p. 70 if necessary.)

3 Two articles in the home which are made of galvanised steel are the dustbin and the cold water tank. Explain why galvanised steel is a suitable metal for each of these uses. Why would steel be unsuitable? What other materials could be used for making (a) dustbins, and (b) cold water tanks?

CROSSWORD ON IRON AND ZINC

First, trace this grid on to a piece of paper (or photocopy this page — teacher, please see the note at the front of the book). Then fill in the answers. Do not write on this page.

Across

1 ___ iron is the purest type of iron (7)

4 In the kitchen you would want yours to be made from stainless steel (5)

7 Arms and legs (5)

10 It is used to protect steel from rusting (5)

11 You can mould articles from ___ iron (4)

13 Chemical symbol for iron (2)

14, 5 down A method of preventing iron from rusting (11, 10)

15 The native state of a metal (4)

16 ___ steel is used to make ball bearings (8)

17 Tin provides this for iron (5)

18 This is what people do with zinc (9)

Down

1 A '___ of iron' means determination (4)

2 You rarely find metals in this state (3)

3 Do this to iron — but don't lose it when you get cross! (6)

5 See 14 across

6 It is used to make steel stainless (6)

8 He made steel (8)

9 Chemical symbol for an element present in Concorde steel (2)

11 This element is present in all steels (6)

12 Thanks for the tantalum! (2)

13 Country establishment where they have always needed iron tools or steel equipment (4)

LEAD

LEAD IN ANCIENT TIMES

Lead occurs widely as lead sulphide, in the ore called *galena*. This is a dense, shiny grey mineral. It is easy to find in the tips outside disused lead mines in the Lake District. The shiny appearance of the ore must have attracted Bronze Age people. They would find it easy to obtain lead from galena. When they roasted the ore, molten lead was formed.

Once they had obtained lead, Bronze Age people must have been rather disappointed. They could not make hard, sharp-edged tools and weapons from lead. Nor was it ornamental.

Centuries later, the Romans found lead very useful. The metal is soft and easily worked. It reacts with water only very, very slowly. The Romans used lead for making water pipes. The Latin name for lead is *plumbum*. This gave its naming to *plumbing* and also to the Welsh word for lead, *plwm*. Wherever they went, the Romans built forts containing bath houses. They probably introduced the idea of taking baths to Britain. They made the boilers and the water pipes from lead.

Over a period of years, lead reacts with water. Lead salts are formed. These dissolve only slightly in water, but they are soluble enough to cause trouble. Lead salts are poisonous. They make people depressed, slow-moving and unable to concentrate. Lead is no longer used for water pipes. Copper and stainless steel are used instead.

THE EXTRACTION OF LEAD

The ore, galena, is concentrated by the flotation method (see p. 6). Then it is smelted (see p. 7).

USES OF LEAD

Lead has important properties. It is never corroded. With its low melting temperature, 330 °C, it is easily shaped and welded. Most metals become brittle after they have been bent a number of times. Lead does not. Lead can be 'cold worked'. Other metals can only be worked after they have been softened by being heated. The world consumption of lead is 4 million tonnes a year.

Lead sheet is used for roofing important buildings. It needs no maintenance and lasts for centuries.

Lead roofing

Underground electric cables are coated with a protective sheath of lead. The electrical wire is surrounded by insulating material. This is wrapped in a protective tube of lead, which keeps out water. Being flexible, the lead sheath allows the cable to be coiled and installed. Underground telephone cables are protected by lead sheaths in the same way.

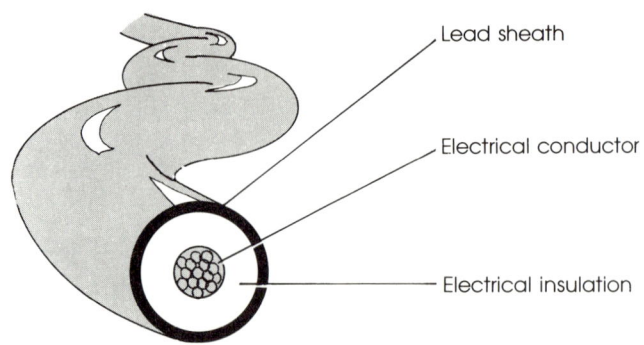

Lead protects electric cable

Lead sheath

Electrical conductor

Electrical insulation

Car batteries are *lead–acid accumulators*. They consist of plates of lead immersed in sulphuric acid. This electrical *cell* produces electricity from the chemical reactions which occur in it.

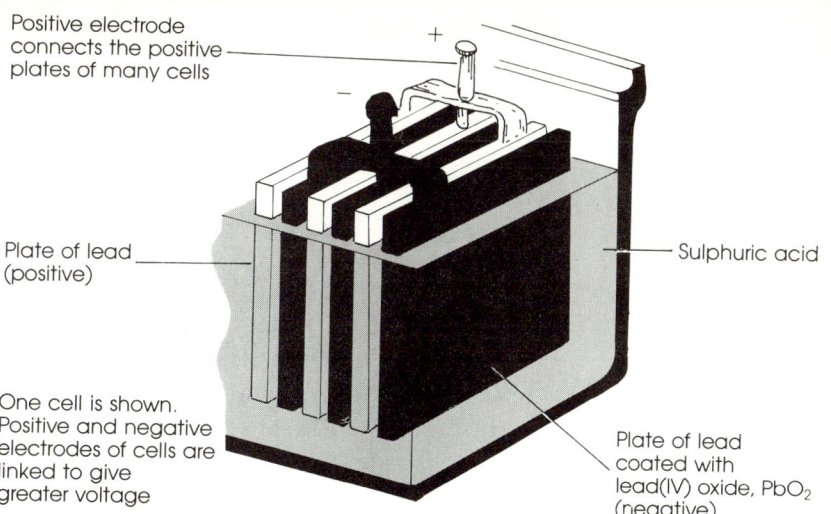

Lead–acid accumulator

Positive electrode connects the positive plates of many cells

Plate of lead (positive)

Sulphuric acid

One cell is shown. Positive and negative electrodes of cells are linked to give greater voltage

Plate of lead coated with lead(IV) oxide, PbO_2 (negative)

Some minor uses of lead are the weighting of yacht keels and the manufacture of lead shot and fishing weights. Recently, hundreds of swans died mysteriously. Post-mortems were carried out. The cause of death was found to be lead poisoning. The birds had swallowed lead weights which had fallen off fishing lines. This is why many fisher-men are now looking for other materials to weight their lines.

Radioactivity cannot pass through thick sheets of lead. Radioactive materials are kept in lead containers. Often in laboratories and hospitals, a wall of lead bricks is built to shield workers from radioactive materials. When handling

A lead apron, used as a shield against radioactivity

radioactive materials, workers often wear lead gloves and lead aprons (see p. 47). Lead is also used as a protection from X-rays.

ALLOYS OF LEAD

A solder is used for joining metals together. It is melted and allowed to drip on to the metals to be joined. When the solder cools and solidifies, it makes a join. Alloys of tin and lead have lower melting temperatures than either of the pure metals. They are used as solders.

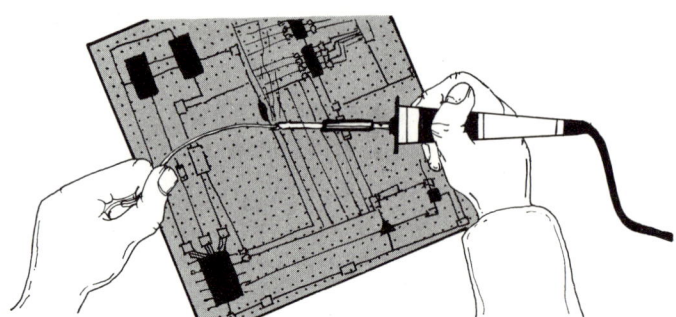

Soldering

EXPERIMENT 7

The lead–acid accumulator

You need two thin sheets of lead, 2 cm × 8 cm, a 100 cm³ beaker, a 4.5 V battery, sulphuric acid and sandpaper, a 1.25 V bulb in a holder and two electrical wires with crocodile clips at the ends.

1) Dip the lead sheets into sodium hydroxide solution. Rinse and dry.

2) Fold the lead strips over the rim of a 100 cm³ beaker. They must not touch each other.

3) Fill the beaker two-thirds full with dilute sulphuric acid.

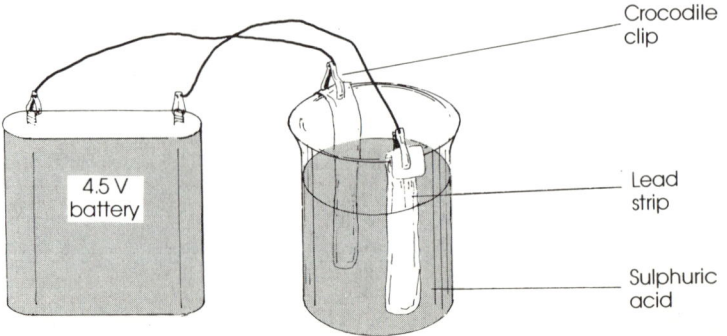

Crocodile clip

4.5 V battery

Lead strip

Sulphuric acid

4) Grip the lead strips with the crocodile clips on the ends of the two electrical wires. Connect them to a 4.5 V battery. Pass a current. Watch for the changes that take place at the surface of the lead strips.

5) After 3 minutes, disconnect the battery. Connect the leads to a 1.25 V bulb. Find out how long it remains lit.

6) The lead strips and sulphuric acid are acting as a battery. If you have a voltmeter, measure the voltage of the battery.

7) When the bulb goes out, recharge the battery as before. Find out whether charging for a longer time affects the time for which the bulb will remain lit.

8) Find out whether the concentration of sulphuric acid has any effect on the ability of the battery to hold a charge.

QUESTIONS ON CHAPTER 7

1 Lead is one of the less reactive metals. This is why it is easy to obtain lead from its ore, lead sulphide. How did primitive people do this?

2 Why did the Romans use lead for making water pipes? Why does lead have the chemical symbol Pb? Why do we not make water pipes from lead now?

3 Since it does not react with water, lead can be used as a protective covering. Give two examples of such use.

4 Which of the solders, A, B and C, would you use for each of the jobs 1, 2 and 3?

Solders
A is a liquid above 180 °C
B melts gradually from 180 °C–230 °C
C has a very high melting temperature

Jobs
1 Mending car radiators
2 Joining wires in an electrical circuit
3 Filling gaps in car bodies

TIN

THE NATURE OF TIN

Tin is a soft, silvery white metal. It is because of its lack of mechanical strength that pure tin is not widely used. Tin is resistant to chemical attack. The largest use of tin is as a protective coating for iron and steel. If the tin coating is scratched, the iron underneath rusts.

CANNED FOOD

In 1810, two scientists called John Hall and Brian Donkin started to do experiments on preserving food. They sterilised food by a heat treatment. They then sealed it into airtight packages. They worked on the possibility of using iron containers. In 1813, the British Army and Navy became interested in the idea. Donkin sent the Duke of Wellington some beef that had been packed in a tinplated iron can. The Duke tried it, and found that it made an appetising meal.

In 1819, Captain Parry set out on a voyage to the Antarctic. He laid in a store of canned food, and found it invaluable. Previous explorers had to rely on hunting for food. Some took a pack of dogs, which they killed, one by one, and ate. One of Parry's ships, HMS Fury, ran aground on the ice and was lost in 1825. Another expedition, led by Captain Ross, found some cans of food in 1833. They ate the food without any ill effects. Two cans brought back by Captain Parry were kept until 1938. When they were opened, after one hundred and thirteen years, the contents were still in perfect condition.

Research workers in the food industry have perfected methods of sterilising food before it is canned. It is very rare indeed to hear of canned food being bad. There is a national outcry with headlines in the newspapers if a can of bad food is opened. We take for granted the achievements of the tinplating industry and the food-processing industry.

TIN PLAGUE

When Captain Scott made an expedition to the South Pole in 1910–12, his party met all kinds of bad luck. They relied less than previous expeditions on dog sleighs and more on motor vehicles. Although they took plenty of petrol, the petrol cans developed mysterious leaks, and they lost some of their petrol. The cans developed holes in the seams. The solder at the seams crumbled. Amundsen, the Norwegian who reached the South Pole a few weeks before Scott, had frequently to resolder his petrol cans.

Captain Scott in Antarctica

These two explorers were both victims of 'tin plague'. Tin is a component of solder. At the time of their expeditions, it was already known that at temperatures below −13 °C tin turns into a grey powder. The grey powder is another form of the element, an *allotrope*. It does not have the strength of white, metallic tin. The lower the temperature falls below −13 °C, the faster the change takes place.

Napoleon was a great respecter of science. He financed research work on aluminium (see Chapter 10). He was one of the first military men to realise the value of canned food to an army on the march. It is a strange coincidence that he too suffered from tin plague. When he and his troops marched deep into Russia in 1812, they were unprepared for the cruel Russian winter weather. One of the difficulties they met was that the buttons on the soldiers' uniforms crumbled. You can guess what they were made of!

QUESTIONS ON CHAPTER 8

1 How long can the food in cans remain edible? What can go wrong with canned food?

2 How did chemistry help an Antarctic exploration? How did a lack of chemical knowledge make life difficult for Antarctic explorers? Who else suffered from unexpected changes at low temperatures?

3 The Duke of Wellington was a famous general. Why did two British scientists think he would be interested in methods of preserving food?

CHAPTER 9

MERCURY

MERCURY IN ANCIENT TIMES

A prehistoric man saw a red rock. He found that, if he dipped it into water, he could use it to draw on the walls of his cave. The rock contained what we call mercury(II) sulphide. Miners call ores containing this mineral *cinnabar*.

The method of extracting mercury from mercury(II) sulphide is very easy. If the ore is crushed and roasted, mercury vaporises. When the vapour is cooled, the liquid mercury that condenses is 99.9% pure. Prehistoric people must soon have found out how to do this. Imagine their surprise when they roasted a red rock and obtained a silvery liquid!

Cave drawings in cinnabar

Mercury is a fascinating metal. It is the only metallic element which is a liquid at room temperature. People found that if they held a pool of mercury in the palm of the hand it moved constantly to and fro. It seemed to be alive, and it was named *quicksilver*. (Quick is an old word meaning *alive*.)

What makes quicksilver move is the slight tremor of the hand caused by blood pumping through the arteries and veins. Please do *not* try holding it in *your* hand all the same! The vapour is poisonous. Another name for mercury was *hydrargyrum.* This is Latin for *silver water.* The chemical symbol for mercury, Hg, comes from this name.

Mercury is 13.5 times as dense as water. It is 1.2 times as dense as lead. It remains liquid down to −39 °C, and boils at 357 °C. As you can see, this range makes it ideal for thermometers.

MINING MERCURY IS NO FUN

The mines in the town of Almaden in Spain produce about half of the world's supply of mercury. They started production in 600 BC. Mining mercury is unhealthy. In the early days, slaves did the job. Later, convicts took over. At one time, the Spanish government allowed men to work in the mines for two years instead of serving in the army. The miners at Almaden now work eight days a month. They hold a second job in addition. If a miner develops 'the shakes' through breathing mercury vapour, he goes to the mine hospital. There he is treated by sitting in a 'hot box' to sweat the mercury out of his system. Some, but not all, of the mercury he has breathed in can be expelled in this way.

As well as miners, others have suffered from mercury poisoning. Laboratory technicians, thermometer-fillers and munitions workers have been affected by mercury poisoning. People who have breathed in too much mercury shake. They cannot walk properly. Their speech is slurred and difficult to understand. They suffer from headaches, sickness and giddiness.

USES OF MERCURY AND ITS COMPOUNDS

With all these dangers, you may wonder why we do not leave mercury underground. The reason is that there is nothing quite like it. Being a liquid metal, mercury can be used as a fluid conductor of electricity. It is used in electrical switches and control devices. The electrical contacts in some thermostats are mercury. Mercury is used to conduct electricity through brine in the manufacture of the important chemicals chlorine and sodium hydroxide.

Mercury is used in fluorescent lights and street lights. It is used in thermometers to measure temperature and in barometers to measure pressure.

Long life mercury batteries provide power for military radios and emergency lights in spaceships. The pacemakers that have been sewn into the hearts of hundreds of thousands of patients to keep their hearts beating regularly are powered by mercury batteries. Long life mercury batteries are also used in hearing aids and cameras.

Ten thousand tonnes of mercury are produced each year. Three thousand different uses are found for the metal and its compounds. Mercury compounds kill bacteria. They are added to paints, floor polishes, paper, and wallpaper paste. They are added to grain before it is stored to prevent the growth of fungi and moulds.

Uses of mercury

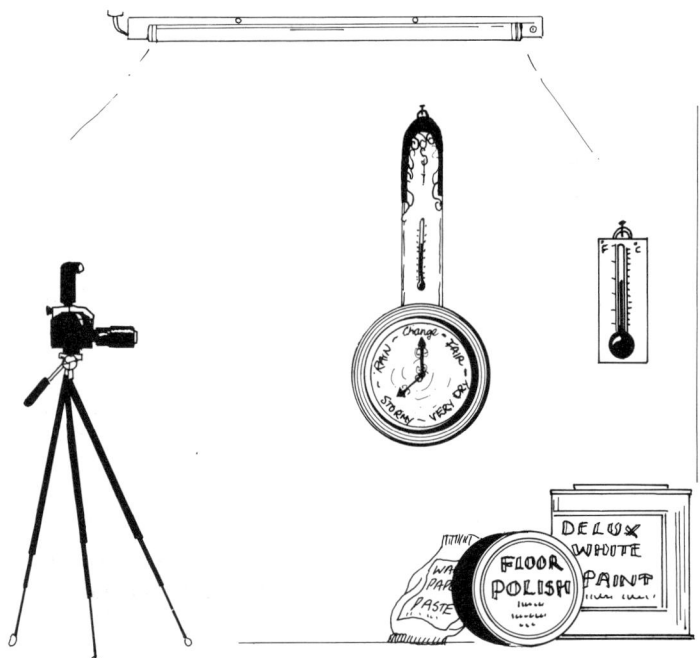

Quicksilver for quick fillings

One use of mercury which benefits a large number of us is in the making of dental *amalgams* for filling cavities in teeth. An amalgam is an alloy of mercury with other metals. Before 1826, gold was the only metal used for filling teeth. Decayed matter was cleaned out of the tooth, and then gold leaf was pressed into the cavity. In 1826, the first dental amalgam was made. Any metal which is used to fill a tooth must be poured as a liquid into the cavity. Once inside, it must start to solidify within seconds. The solid filling must be hard enough to bite on without cracking. It should not become tarnished. It must not react with water or with food and drink.

Mercury is the only metal which is a liquid at room temperature. Dental amalgam is made by shaking a powdered mixture of silver, tin, copper and zinc with mercury. Within seconds, the metals dissolve to form an amalgam, with which the dentist packs the cavity. In five minutes, the amalgam sets to form a solid. One or two hours later, the solid will have become hard enough to chew on. The different metals in the amalgam are there for different reasons. Tin bonds to mercury, and helps the amalgam to set. Silver helps the filling to stay free from tarnish. It also slows down the hardening process, giving the dentist time to mould the filling into the cavity. Copper and zinc give the alloy strength.

Dental amalgam does not produce mercury vapour. This means that you cannot breathe in mercury from your fillings! Dental amalgam does not react with any of the substances which find their way into your mouth. For this reason, no soluble mercury compounds can be formed. This means that you cannot swallow the mercury in your fillings!

Dental amalgam

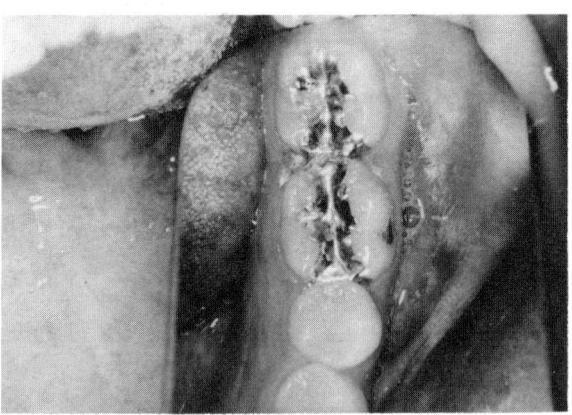

THE NASTY SIDE OF MERCURY

This beautiful, useful, silvery liquid has a darker side to its nature. Minamata is a small fishing village in Japan. In 1953, a strange illness began to affect the people of Minamata. They became tired and irritable. They complained of headaches, blurred vision and poor hearing. Their arms and legs felt numb. They had difficulty in balancing, and had fits of shaking. Doctors had not seen these symptoms before. The cause of the 'Minamata disease' was a mystery.

The doctors noticed that almost all the patients were from fishermen's families. Scientists analysed fish caught in Minamata Bay. They found that these fish contained a very high level of mercury. By 1963, they had proved that this mercury caused Minamata disease.

Where did the mercury come from? Two years before the epidemic started, a plastics plant was built on Minamata Bay. Each year, the plant discharged a small quantity of mercury compounds into the Minamata River. The level of mercury in the water was 2 parts per billion (ppb). (1 billion = 1 thousand million.) Although this level of mercury is higher than in the middle of the sea, it is safe to drink. What the manufacturers did not know was that a food chain was being set up (see the illustration below).

A food chain

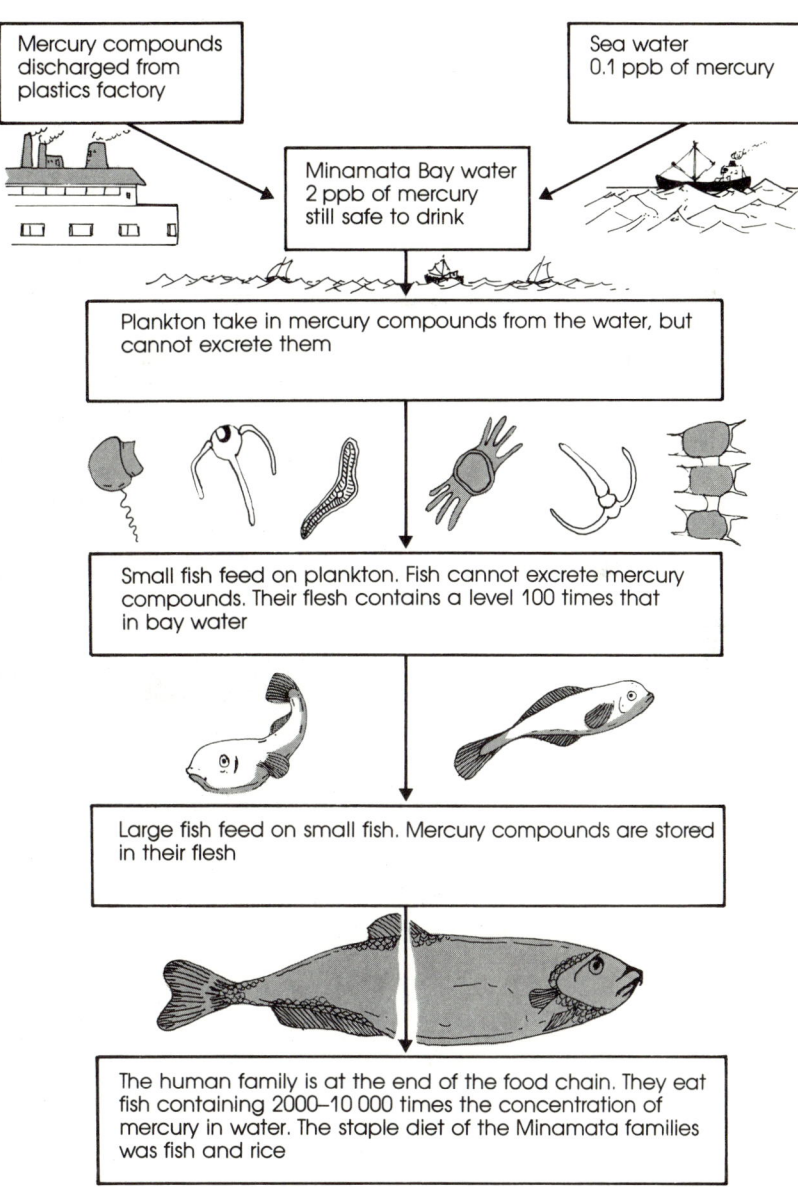

Mercury compounds discharged from plastics factory

Sea water 0.1 ppb of mercury

Minamata Bay water 2 ppb of mercury still safe to drink

Plankton take in mercury compounds from the water, but cannot excrete them

Small fish feed on plankton. Fish cannot excrete mercury compounds. Their flesh contains a level 100 times that in bay water

Large fish feed on small fish. Mercury compounds are stored in their flesh

The human family is at the end of the food chain. They eat fish containing 2000–10 000 times the concentration of mercury in water. The staple diet of the Minamata families was fish and rice

By the time the cause of the epidemic has been found, 43 people had died. The chemical company stopped pouring mercury compounds into the bay. Later, the number of deaths reached 100. Hundreds of people were seriously harmed. Many are still helpless and blind. In all, 10 000 people were affected by mercury poisoning.

Both mercury and mercury compounds are poisonous. The 'chlor-alkali' plants, which make chlorine and sodium hydroxide, use mercury. For a long time, people discharged mercury into lakes and rivers. Since the Minamata tragedy, they have been more careful about this. The mercury already in the lakes and rivers will go on polluting our environment for years to come. It is one of the most long-lasting of pollutants.

Mercury is a marvellous metal. It helps us in thousands of ways. There is no other liquid metal, and nothing else can take its place. In looking after our needs, we have poured mercury into the environment. It has dealt out disease and death. The mercury tragedies at Minamata and elsewhere show the difficulties which scientists face. We want them to provide for our needs and our comfort. At the same time, they have to avoid polluting the environment.

QUESTIONS ON CHAPTER 9

1 Why is mercury called *quicksilver?* Why does it have the chemical symbol Hg? How is it possible to obtain quicksilver from a red rock?

2 'There is nothing quite like mercury.' Explain this statement.

3 Is there any mercury in your body? Where is it, and why is it there? Why is mercury the metal that is chosen for this purpose?

4 List four uses of mercury.

5 Explain what is meant by a *food chain.* How was a food chain important in the Minamata story?

6 Coal has an average mercury content of 0.2 ppm (parts per million). A power plant burns 1 million tonnes of coal in a year. What mass of mercury is sent into the air by a single power plant?

ALUMINIUM

THE NEWCOMER

Copper has been known since 5000 BC, bronze since 3500 BC, and iron since 1200 BC. Aluminium is a newcomer on the scene. It was first obtained in AD 1827.

The extraction of aluminium from its ore, *bauxite*, was a difficult problem. Aluminium is a very reactive metal. Reactive metals are difficult to extract from their compounds. A French chemist called Henri St Claire Deville succeeded in producing aluminium on a small scale in 1854. He made an aluminium medal for his emperor, Napoleon Bonaparte III. The emperor gave money for experiments on aluminium. He wanted to manufacture the metal on a large scale. He was impressed with its low density. He also liked the fact that it was not corroded by air or water. He could see what an advantage it would be for his troops to wear aluminium helmets and breastplates. While the metal cost £200 a kilogram, his dream could not come true.

In 1886, the problem of extracting aluminium on a large scale was solved. A Frenchman called Paul Héroult and an American called Charles Hall had the same idea at the same time. Both men were 22 years old when they made their discovery. Both were working in makeshift laboratories. Hall worked in an outhouse of his father's vicarage. Héroult worked in a back room in a local tannery. Hall and Héroult used electricity to extract aluminium. Some compounds can be *electrolysed*. This means that they can be split up by passing electricity through them. Such compounds must be melted or dissolved before they will conduct electricity. People had tried to electrolyse aluminium oxide before. The problem was its high melting temperature, 2050 °C. They found it impossible to keep aluminium oxide molten while they experimented with it. Charles Hall melted aluminium sodium fluoride. This is mined as an ore called *cryolite*. He dissolved aluminium oxide in the melt. He was able to keep it molten while he passed an electric current through it for some minutes. To his delight, he found small beads of aluminium. Four thousand miles away, Paul Héroult had arrived at the same solution to the aluminium problem.

The plant used for the extraction of aluminium is called a Hall–Héroult cell (see below). It can operate non-stop. Aluminium oxide is fed in at the top. Molten aluminium is tapped off at the bottom.

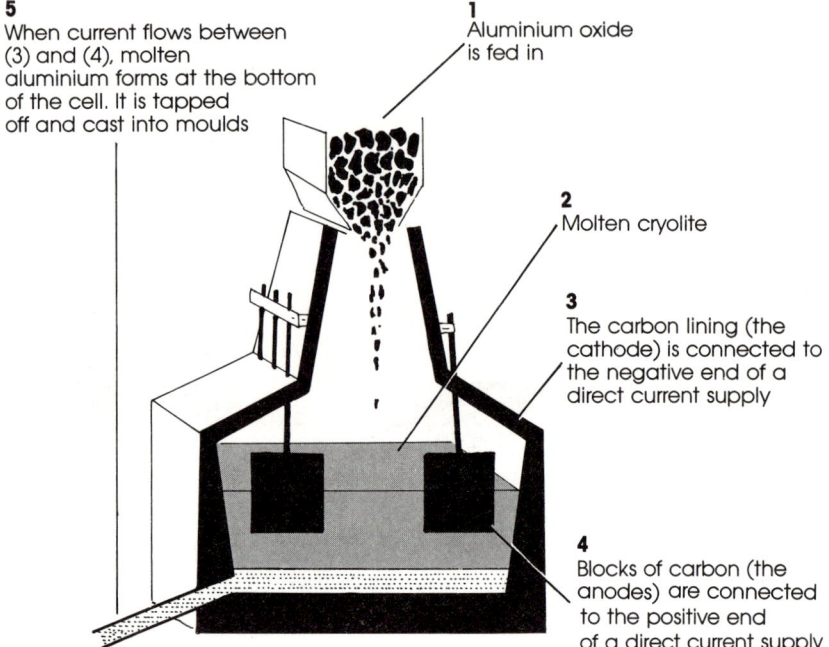

5
When current flows between (3) and (4), molten aluminium forms at the bottom of the cell. It is tapped off and cast into moulds

1
Aluminium oxide is fed in

2
Molten cryolite

3
The carbon lining (the cathode) is connected to the negative end of a direct current supply

4
Blocks of carbon (the anodes) are connected to the positive end of a direct current supply

The Hall–Héroult cell for obtaining aluminium

In 1880, aluminium cost £220 000 a tonne. After the Hall–Héroult process started production, the price fell dramatically. In 1910, the price was £85 a tonne. The process uses a great deal of electricity. To produce one tonne of aluminium takes 15 000 kilowatt hours of electricity. This would keep a two-bar electric fire burning day and night for one year. Aluminium plants are constructed in regions which have low cost electricity. This usually means hydro-electric power, electricity from power stations driven by waterfalls. Hydroelectric power stations are common in Norway and Canada. They are also to be seen in the Highlands of Scotland. Aluminium oxide is transported to a plant near the power station at Loch Leven in a very beautiful part of Scotland. Hydroelectric power stations are often in areas of great natural beauty, but the presence of aluminium smelters is unsightly. They often spoil the natural beauty of the region. This is a problem. Conservationists want to preserve the countryside. Industrialists want to be able to obtain more and more aluminium. As you will see, this marvellous metal can do an amazing variety of jobs.

USES OF ALUMINIUM

Aluminium is a very reactive metal. It is more reactive than iron, and will displace iron from iron oxide:

Aluminium + Iron oxide → Iron + Aluminium Oxide

As aluminium and oxygen combine, a large quantity of heat is given out. The reaction is called the *thermit* reaction. (*Therm* means heat in Latin.) As a result, the iron which forms is molten. This makes the thermit reaction useful for welding jobs. It can be used to mend a break in a railway line. The gap is filled with iron oxide, aluminium powder and a fuse. The fuse is lit. This sets off the thermit reaction. A plug of molten iron forms and welds the ends of the broken rail together.

The thermit process in action

Break in the line

Pack with iron oxide and aluminium powder. Insert fuse. Light and draw back

Admire welding job

How does aluminium stay as good as new?

When a fresh surface of aluminium meets the air, it rapidly combines with oxygen in the air. The metal becomes coated with a layer of aluminium oxide. This layer protects the metal beneath it from chemical attack. With its protective coating, aluminium gives the impression of being an unreactive metal. The metal that reacts so vigorously in the thermit reaction will not react further with air or water or with cold acids and alkalis. Aluminium is never corroded under normal weather conditions. There are chemicals that will remove the oxide layer. When this is done, aluminium shows its true reactivity.

A process called *anodising* is used to make the surface layer of aluminium oxide thicker. Aluminium is made the positive electrode (the anode) in an electrical circuit. The anodised aluminium is able to absorb dyes. The result is an attractively coloured metal. It is used for making doors and window frames.

This statue of Eros is made of aluminium. It has stood in Piccadilly Circus for 100 years without being corroded

Aluminium is completely non-toxic. There is no danger of it contaminating food. It can safely be used for milk bottle tops, for food containers and for baking foil.

Recycling aluminium

Since aluminium never becomes corroded, scrap aluminium can be recycled. It needs only to be melted and reused. The aluminium foil used for cooking, for milk bottle tops and for food containers is all as good as new. Collecting used aluminium and recycling it is the only sensible thing to do. An enormous amount of aluminium is thrown away, but some recycling does take place. Oxfam shops collect aluminium for recycling. The price that Oxfam obtains for it depends on its purity. During collection, they must take care not to include other metals. Twenty tonnes of aluminium can be recycled with the amount of electricity used to make one tonne of new metal.

Aluminium is used to reflect light and heat

Aluminium reflects light, and is used for making the reflectors in car headlights. It also reflects heat, and is used as an insulator. A thin film of aluminium can be bonded to polyester fabric. This makes an insulating blanket. Premature babies are often wrapped in these aluminium blankets to keep them warm. Aluminium reflects the body's heat, and acts as an insulator. Firefighters wear asbestos suits coated with aluminium. The aluminium reflects the heat of the fire, and keeps the wearer cool. At least, he is cooler than he would be without it! The ability of aluminium to reflect heat keeps the baby warm and keeps the firefighter cool. Farmers find that cows give more milk when cool.

In hot countries, aluminium sheets are put on to the roofs of cowsheds. These roofs reflect heat and keep the cows cool.

(a)   (b)

(a) An insulating blanket

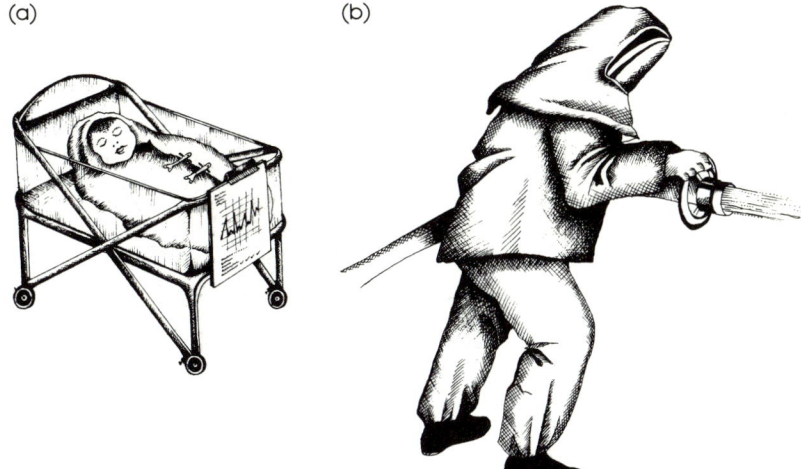

(b) A firefighter's suit

Aluminium is used to conduct heat

When it is in direct contact with a heat source, aluminium conducts heat. For saucepans, baking tins and frying pans, aluminium is widely used. We use 10 000 tonnes of aluminium foil a year for baking and wrapping food.

Aluminium is used to conduct electricity

Aluminium is replacing copper in overhead transmission lines. Aluminium is as good an electrical conductor as copper and only one-third as dense. Since aluminium cables are lighter than copper cables, the pylons which support them can be less sturdily built, with a saving in cost.

ALUMINIUM ALLOYS

Aluminium has a low density. It is the least dense of the common metals. It is only 2.7 times as dense as water. Aluminium resists corrosion. These two characteristics give aluminium advantages over other metals for many purposes. It has one drawback. Pure aluminium is a soft metal, easily cut by iron. When small quantities of other elements are added, the metal becomes stronger. One alloy of aluminium is *Duralumin* (*dur* is Latin for 'hard'). It contains 5% copper and small quantities of other elements. It has the strength of mild steel and has only one-third of the density. It will stand up to bad weather conditions without being corroded. This alloy is ideal for the manufacture of aeroplanes.

An aeroplane made of aluminium alloys

Another alloy of aluminium and titanium is ideal for high-speed aircraft. As well as being strong and of low density, it has a high melting temperature. This is important for an aeroplane which flies at high speed. As it cuts through the air friction may raise the temperature of the outside of the aircraft to 150 °C. The metal must not soften.

Aluminium alloys are ideal for the construction of boats. Yachts and many other small craft are made of aluminium. They are light, and they resist corrosion. Ocean liners and tankers need the strength of steel. Although battleships have steel hulls, their superstructures are made of aluminium alloys. The bridge, the radar scanners and the other structures you can see above deck are made from aluminium alloys. This makes it possible to build up above sea level without making the ship top-heavy. It has a disadvantage, however, which came to light during the Falklands War. Aluminium is a very reactive metal. If an enemy missile hits the aluminium superstructure, the aluminium burns fiercely. The fire spreads so fast that it cannot be extinguished. Battleships were destroyed during the Falklands War by a single hit because of the speed with which aluminium burns. What is the solution to this problem? I cannot tell you. Scientists and naval experts are working on it.

Aluminium alloys are used in the construction of this ship

ALUMINIUM AND IRON: PARTNERS IN THE CAR INDUSTRY

Aluminium costs six times as much as mild steel. Despite this, more and more parts of motor vehicle engines are being made of aluminium alloys. Rocker arm covers, radiator grilles, pistons, distributor caps, trim and engine blocks can be made from aluminium alloys. As more of the vehicle is made of aluminium, it becomes lighter. Lighter vehicles do more miles to the gallon. The saving in petrol is important. We are burning too much of the world's supply of petroleum oil. We should be turning it into petrochemicals instead of burning it. (Plastics, dyes, drugs, paints and solvents are petrochemicals.) For some parts of a vehicle, the strength of steel is needed. The pistons and cylinder linings must be of steel. Land Rovers and Range Rovers have aluminium bodywork. As you know, a big problem with cars and other vehicles is rusting of the bodywork. Aluminium never rusts. When the steel chassis of a Land Rover eventually rusts, the aluminium body parts are sound. They should be recycled. This does not happen. We are too wasteful of materials. We ought to be much more careful in our use of the Earth's resources.

EXPERIMENT 8

Anodising and dyeing aluminium

You need aluminium foil, sodium hydroxide solution, sulphuric acid and a 4.5 V battery.

1) Cut pieces of aluminium foil, one 4 cm × 12 cm and the other 20 cm × 12 cm.

2) Clean the two pieces of foil. Soak them in sodium hydroxide solution for 10 minutes (this is important). Rinse them with water. Dip them into nitric acid. Rinse with water.

3) Bend the larger piece of foil into a cylindrical shape. Attach crocodile clips to both pieces.

4) Assemble the apparatus shown below.

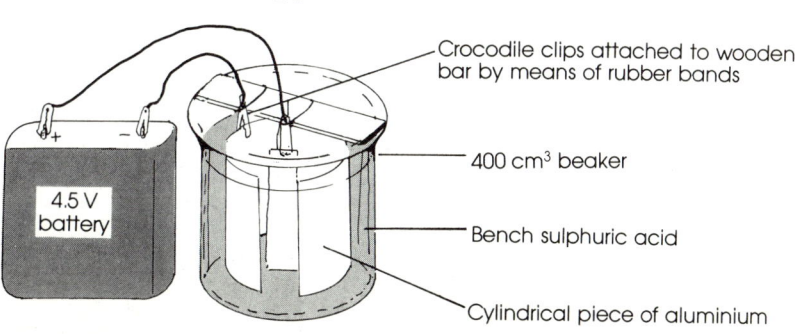

Anodising aluminium

Crocodile clips attached to wooden bar by means of rubber bands

400 cm³ beaker

4.5 V battery

Bench sulphuric acid

Cylindrical piece of aluminium

5) The cylinder and the smaller strip of aluminium must not touch. The cylinder is connected to the negative end of the battery. The strip is connected to the positive end of the battery. This makes the strip the *anode* in the cell. Pass a current for 20 minutes.

6) Make a solution of the dye alizarin (or eosin or Congo red). Heat the dye solution almost to boiling temperature. Immerse the anodised strip of aluminium for 5 minutes. Rinse.

Note: The experiment works better with aluminium thicker than baking foil.

QUESTIONS ON CHAPTER 10

1 Aluminium is used for the manufacture of (a) saucepans, (b) overhead cables, (c) boats, (d) aeroplanes, (e) hospital blankets, (f) milk bottle tops, (g) firefighters' suits. Explain what property of aluminium makes it suitable for each of these uses.

2 Why was Napoleon interested in aluminium?

3 What did a young Frenchman and a young American do to reduce the price of aluminium? Why is aluminium still dearer to produce than steel?

4 The thermit reaction shows that aluminium is a reactive metal.
Aluminium is used for doors and window-frames.
How can both of these statements be true?

5 Aluminium is used as a conductor of heat.
Aluminium is used as a thermal insulator.
How can both of these statements be true?

6 Explain why recycling scrap aluminium is easier than recycling scrap iron.

CROSSWORD ON ALUMINIUM

First, trace this grid on to a piece of paper (or photocopy this page — teacher, please see the note at the front of the book). Then fill in the answers. Do not write on this page.

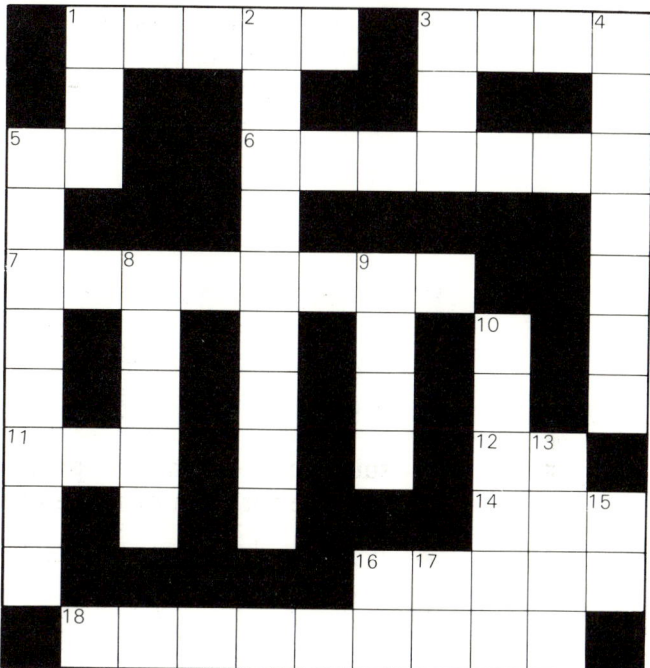

Across

1 Make a piece of aluminium an ____ in order to dye it (5)
3 See 9 down
5 Chemical symbol for chromium (2)
6 Some people ____ aluminium so that it can be used again (7)
7 A military man who supported science (8)
11 This may be made of aluminium or iron (3)
12 Chemical symbol for thallium (2)
14 See 10 down
16 You have to heat aluminium oxide to a very high temperature before it ____ (5)
18 Aluminium ____ you from radiant heat (9)

Down

1 This reacts quickly with aluminium (3)
2 An alloy of copper and aluminium (9)
3 I bet Hall gave a ____ of joy when he obtained aluminium (3)
4 In a ____ reaction aluminium combines with oxygen (7)
5 Aluminium ____ electricity (8)
8 You need aluminium to make this flier (5)
9, 3 across Aluminium is obtained from ____ ____ mines (4, 4)
10, 14 across You may find milk under this piece of aluminium (6, 3)
13 It was difficult to get ____ of aluminium before 1886 (4)
15 Postscript (2)
16 Not a Master of *Science* (2)
17 And in France he wanted to go home! (2)

QUESTIONS ON METALS AND ALLOYS

1 Look at the picture on p. 1. Now you have read the book, can you say which metals are used for the manufacture of all the objects shown?

2 The table below lists the metals mentioned in this book. They are ranked in order of reactivity. Number 1 is the most reactive. They are ranked in order of density. Number 1 is the densest.

Look at the picture on p. 1.

Name	Chemical symbol	Position in rank order of chemical reactivity	Position in rank order of density
Aluminium	Al	1	9
Copper	Cu	6	5
Gold	Au	9	1
Iron	Fe	3	6
Lead	Pb	5	3
Mercury	Hg	7	2
Silver	Ag	7	4
Tin	Sn	4	7
Zinc	Zn	2	8

Metals

A Look at the rank order of density.
   (a) Which is the least dense of the metals listed? What is this metal used for? Give examples of uses which need a low density metal.
   (b) Which is the densest metal? What is this metal used for?
   (c) Which metal ranks third in order of density? What uses of this metal depend on its high density? What other uses does it have? What property of the metal makes it suitable for the uses you mention?
   (d) Which metal comes second in order of density? Why is it not used for the same purposes as (c)? How is it used?

B Look at the rank order of reactivity.
   (a) Which is the least reactive of the metals listed? How is this metal found in nature?
   (b) Which is the most reactive of the metals listed? How is this metal obtained from its ore? Why is it that this was the last metal on the list to be obtained as the free metal?
   (c) In order of tonnes used each year, iron comes top of the list. Why is it such a useful metal? Why is it possible to obtain iron from its ores fairly cheaply? In your answer, say where they come from and whether they are expensive materials.

(d) In order of tonnes used each year, aluminium comes second. Why is it such a useful metal? For what purpose is aluminium preferred to iron? Give 3 examples. For what purposes is iron better than aluminium? Give 3 examples.

METALS AND ALLOYS WORDSQUARE

Trace or photocopy this wordsquare — teacher, please see the note at the front of the book. The names of 15 metals and alloys are hidden in it. Two occur twice. Circle each name. The words read diagonally as well as vertically and horizontally. Do not write on this page.

R	D	L	O	G	N	O	R	I	M
E	B	W	R	O	U	G	H	T	E
D	U	R	A	L	U	M	I	N	R
L	E	S	A	D	R	M	X	E	C
O	M	B	C	S	U	A	P	L	U
S	U	N	R	I	S	P	A	E	R
Z	I	N	C	O	O	T	S	E	Y
Z	R	L	D	C	N	R	O	T	R
T	A	S	O	D	I	U	M	S	S
C	B	C	A	S	Y	L	E	A	D

ANSWERS TO WORDFINDER ON P. 9

1 Pan
2 Prospector
3 Ductile
4 Wire
5 Mercury
6 Molybdenum
7 Sodium
8 Open-cast
9 Break-even
10 Flotation
11 Grind
12 Native
13 Slag
14 Gangue
15 Alloy
16 Solder
17 Lead
18 Tin
19 Iron
20 Carbon
21 Cornwall

Message: Well done! You have discovered the gold!

EXPERIMENT 2 (p. 15)

Copper sulphate floats off the froth. Sand remains in the beaker.

EXPERIMENT 6 (p. 39)

The untreated nail rusts quickly in sodium chloride solution. The galvanised nail does not rust. Even the part which is not covered with zinc does not rust. This is because zinc is a more reactive metal than iron. As long as zinc is present, zinc is corroded while iron remains untouched.

QUESTIONS ON CHAPTER 6 (p. 43)

Question 2 Hydrogen is formed by the reaction between zinc and steam. It can cause an explosion, especially in a closed container.

QUESTIONS ON CHAPTER 7 (p. 49)

Question 4 A2, B3, C1

QUESTIONS ON CHAPTER 9 (p. 58)

Question 6 0.2 tonne

ANSWERS TO WORDSQUARE ON p. 69

barium, gold (2), Duralumin, wrought iron, cast iron, lead, mercury, solder, bronze, tin, copper, steel, calcium, zinc (2), brass

METALS AND ALLOYS IN THE PICTURE ON p. 1

aluminium alloys: aeroplane, window frame, baking tin, saucepan
brass: scales, warming pan
copper: kettle
tin-plated steel: food cans
chromium-plated steel: washing machine door, wastebin lid, legs of stool, bicycle handlebars
wrought iron: gate
galvanised steel: girders, garage door
mercury: in thermometer
gold: earrings
silver: necklace
steel: car, bicycle

INDEX

Aircraft 64
Alloy 3, 12, 13, 14, 20, 21, 35, 48, 63, 64
Aluminium 7, 8, 59–67
    alloys 63–4
    uses 61–3
Amalgam 55
Amundsen 51
Andes 19
Annealing 28, 38
Anodising 61, 65
Antarctic 50
Atahuallpa 20
Australia 21

Bauxite 59
Bessemer converter 32, 33
Blacksmith 30
Blast furnace 31, 32
Brass 10, 15, 51
Break-even point 7, 21
Bronze 10, 12, 14, 20, 28, 45
Bronze Age 12, 27

Car industry 65
Carat 21
Carbon 7, 11, 27, 30, 31, 35
Cast iron 27, 30, 31
Charcoal 11, 27
Chromium 35, 37
Cinnabar 53
Circuit board 16
Coins 15, 21, 24
Coke 7, 31
Concentrating ore 6
Copper 8, 10–22, 45
    compounds 13
    conductor 14
    pyrites 13
    uses 13, 56
Crusher 4, 5

Donkin 50
Ductility 2
Duralumin 63

Egypt 18
Electrolysis 60
Electroplating 42
Element 2
Eros 62
Etching 16

Falklands 64
Flotation 6, 15, 45
Food chain 57
Food industry 50
Furnace 11, 12, 27

Galena 45
Galvanising 37, 41
Gauls 29
Gold 7, 8, 18–22
    alloys 21
    leaf 21
    plating 22
    reserves 21
    rolled 22
    rush 21
    from seawater 21
    uses 21, 22, 25, 55

Hall 50, 59
Hallmark 24
Heat treatment of iron and steel 27–9, 38
Héroult 59
Hopi 23
Hydroelectric power 60

Incas 19
Industrial Revolution 31
Iron 8, 27–40
Iron Age 20, 27

Japanese 29
Jewellery 21, 23

Land Rover   42, 65
Lead   8, 45-9
    accumulator   46-8
    alloys   48
    apron   47
    roofs   46
    salts   45
    uses   45
Limestone   31

Malleability   2
Mercury   53-8
    mining   54
    uses   54
Metallurgist   3
Metallurgy   3, 29
Minamata   56-8
Minerals   3, 4
Mining   4, 5
Molybdenum   3

Napoleon   51, 59
Native metals   8
Navaho   23
Nickel   35, 37

Ores   3, 7

Panning   4, 21
Parry   50
Peru   19, 20
Photocells   24
Plumbing   45
Pollution   58
Properties of metals   2
Prospecting   3

Quenching   28, 29, 38
Quicksilver   53-5

Radioactivity   47
Reactive metals   7, 8
Recycling   62, 65
Reduction   7
Romans   28, 45
Ross   50
Rust-proofing   36

Sacrificial protection   43
St Claire Deville   59
Samurai   29
Scotland   60
Scott   51
Screening   6
Sherardising   42
Ships   43, 64
Silver   23-6
    coins   24
    mirrors   24
    plating   24
    uses   23, 56
Slag   7, 32
Smelting   7, 11, 45
Solar collectors   24
Solder   48
South Africa   21
South America   19
Space shuttle   25
Spaniards   20
Steel   32-40
Stone Age   10, 12, 18
Sulphur   36
Sulphur dioxide   36
Swords   28

Tempering   28, 38
Thermit reaction   61
Tin   12, 50-2
    plague   51
    uses   48
Titanium   35, 64
Tungsten   35
Tutankhamun   18

Vikings   28

Weathering   13, 16
Wellington   50
Wrought iron   27, 30

X-rays   48

Zinc   37, 38, 41-4
    uses   41, 56
Zuni   23